Nematodes - Recent Advances, Management and New Perspectives

Edited by Cristiano Bellé and Tiago Edu Kaspary

Published in London, United Kingdom

IntechOpen

Supporting open minds since 2005

Nematodes - Recent Advances, Management and New Perspectives
http://dx.doi.org/10.5772/intechopen.94719
Edited by Cristiano Bellé and Tiago Edu Kaspary

Contributors
Amar Bahadur, Tabassum Ara Khanum, Nasira Khatoon, Nasir Mehmood, Ricardo R. Balardin, Daiane Dalla Nora, Rodrigo Ferraz F. Ramos, Zaida I. Antoniolli, Cristiano Bellé, José Carlos V. Rodrigues, Aasha Rana, Ashok K. Kumar Chaubey, Aashaq H. Hussain Bhat, Himani Sharma, Ali Anwar, Efath Shahnaz, Saba Banday, Mohad. Mughal, Taibah Bashir, Qadrul Nisa, Gulam Jeelani, Ahmed Nasri, Patricia Aïssa, Hamouda Beyrem, Ezzeddine Mahmoudi, Saroj Yadav, Jaydeep A. Patil, Ivana Majić, Gabriella Kanižai Šarić, Ankica Sarajlić, Emilija Raspudić, Marko Josipović, César Ángel-Sahagún, Oscar Barrón-Bravo, Ismael Montiel-Maya, Ana Cruz-Avalos, Fidel Avila-Ramos, Jaime Molina Ochoa, Sunil Kumar, Ranjit Kumar, Pankaj Sood, Nishi Keshari, Gurram Mallikarjun, Linnley Mulusa

First published in London, United Kingdom, 2022 by IntechOpen
IntechOpen is the global imprint of INTECHOPEN LIMITED, registered in England and Wales, registration number: 11086078, 5 Princes Gate Court, London, SW7 2QJ, United Kingdom
Printed in Croatia

British Library Cataloguing-in-Publication Data
A catalogue record for this book is available from the British Library

Additional hard and PDF copies can be obtained from orders@intechopen.com

Nematodes - Recent Advances, Management and New Perspectives
Edited by Cristiano Bellé and Tiago Edu Kaspary
p. cm.
Print ISBN 978-1-83969-650-3
Online ISBN 978-1-83969-651-0
eBook (PDF) ISBN 978-1-83969-652-7

For EU product safety concerns:
IN TECH d.o.o., Prolaz Marije Krucifikse Kozulić 3, 51000 Rijeka, Croatia, info@intechopen.com or visit our website at intechopen.com.

Meet the editors

Cristiano Bellé is a researcher in the field of plant nematology and biological control at Phytus Institute, Brazil. He is also a researcher at the Universidad Nacional de San Agustín de Arequipa (UNSA), Peru. He obtained a master's degree in Agronomy from the Federal University of Santa Maria, Brazil, in 2014, a doctorate degree in Crop Protection from the Federal University of Pelotas, Brazil, in 2018, and a post-doctorate degree in Soil Biology and Microbiology from the Federal University of Santa Maria in 2019. He has experience in the field of agronomy with an emphasis on plant pathology and plant nematology. His work is related to the etiology and diagnosis of plant diseases caused by nematodes and fungi, pathogenicity and host-parasite relationships of plant-parasitic nematode diseases, the interaction between plant-parasitic nematodes and soilborne fungi in plants, and biological, cultural, and chemical control of plant-parasitic nematodes.

Dr. Tiago Kaspary obtained a Ph.D. in Weed Science from Universidade Federal do Rio Grande do Sul (UFRGS), Brazil, and a sandwich Ph.D. from the University of Arkansas, USA. He is currently a weed science researcher at the Instituto Nacional de Investigación Agropecuaria (INIA), La Estanzuela, Uruguay (SNI - I). He has experience in the field of agronomy with an emphasis on weed science. His works are related to biology and weed management, the interaction of weeds with pathogens and nematodes, the resistance of weeds to herbicides, and molecular biology applied to weed science.

Contents

Preface

Nematodes can infect humans, animals, and plants. They can cause serious damage and yield losses in a wide range of crops throughout the world. This imposes a challenge to the sustainable production of food worldwide.

Thus, finding sustainable methods to control these pathogens is very important. In this book, we discuss identification and characterization methods based on molecular and morphological techniques, incidence and impacts on large crops, and nematode parasitizing of alternative species (weeds, ornamentals, invasive plants, and volunteer plants).

We also examine the management of these parasites in an integrated manner, using specific monitoring, chemical and biological control methods, genetic host resistance, and cultural methods of nematode control, taking into account the capacity of these phytoparasites to be affected by the imbalance of productive systems and to comply with the function of environmental bioindicators.

This book discusses the current and emerging challenges of the presence of nematodes in different crop production systems.

Cristiano Bellé, Ph.D.
Phytus Institute,
Santa Maria, Rio Grande do Sul,
Brazil

Tiago Edu Kaspary, Ph.D.
Instituto Nacional de Investigación Agropecuaria del Uruguay,
Montevideo, Uruguay

Section 1

Diagnostic Nematode in Crops

Chapter 1

Nematodes Diseases of Fruits and Vegetables Crops in India

Amar Bahadur

Abstract

Nematodes are the most plentiful animals on earth, commonly found in soil or water, including oceans. Some species of nematodes are parasites of plants and animals. Plant-parasitic nematodes are non-segmented microscopic, eel-like round worms, obligate parasite possess stylets that live in soil causing damage to plants by feeding on roots or plant tissues. Plant-parasitic nematodes feed on roots, either within the root, some nematodes feed leaves. These nematodes cause breakdown of resistance to fungal diseases in fruit crops. Plant-parasitic nematodes living host tissue to feed on to grow and reproduce. Nematode life cycle consists of an egg, 4 pre-adult stages (juveniles) and an adult, life cycle depending on the species and the temperature. Nematodes do not move long distances (less than 6 inches per year). They are usually transported over long distances on machinery, in nursery stock, transplants, seeds, or by animals, moves soil, water and wind. They acquire nutrients from plant tissues by needle-like feeding structure (stylet/spear). Nematodes can be classified into three groups depending on feed on the plants such as ecto-parasitic nematodes are always remaining outside the plant root tissues. Migratory endoparasitic nematodes move through root tissues sedentary endoparasitic nematodes penetrate young roots at or near the growing tip. They steal nutrients, disrupt water and mineral transport, and provide excellent sites for secondary pathogens (fungus and bactria) to invade the roots and decay. Several nematode species that cause problems in fruit orchards that are major limiting factors in fruit crop production cause extensive root necrosis resulting in serious economic losses. The root-knot nematode (*Meloidogyne* spp.*)*, burrowing nematode (*Radopholus similis*) and citrus nematode (*Tylenchulus semipentrans*) are the major nematode pests that infect fruit crops. Parasitic nematodes that can damage tree fruit roots. Many kinds of nematodes have been reported in and around the roots of various fruit crops, only few are cause serious damage, including Root-knot nematodes (*Meloidogyne* spp.), Lesion nematodes (*Pratylenchus* species), Ring nematodes (*Mesocriconema* spp) are cigar-shaped that are strictly ectoparasitic, Dagger nematodes (*Xiphinema* spp) are relatively large ectoparasites that feed near root tips, Sting nematodes (*Belonolaimus* species) are ectoparasitic, Citrus nematodes *(Tylenchulus semipenetrans)* are sedentary semi-endoparasites. Nematodes reduce yield without the production of any noticeable above ground symptoms. Typical above ground symptoms of nematode infections stunting, yellowing and wilting. Major nematodes associated in large number of vegetables crops in India such as root-knot nematodes (*Meloidogyne* spp.), cyst nematodes (*Heterodera* spp.), lesion nematodes (*Pratylenchus* sp.), reniform nematodes (*Rotylenchulus* sp.) lance nematodes (*Hoplolaimus* spp.), stem and bulb nematode (*Ditylenchus* spp.) etc. Root-knot nematodes are important pests of vegetables belonging to solanaceous (brinjal, tomato, chili), cucurbitaceous (biter ground, cucumber, pumpkin, bottle gourd) leguminous (cowpea, bean, pea), cruciferous cauliflower, cabbage, broccoli, brussels, sprout), okra and several other root and bulb crops (onion, garlic, lettuce, celery, carrot, radish). Four species (*M. incognita, M. javanica, M. arenaria and M. hapla*) are more than 95% of the root-knot nematode population worldwide distribution. Stem and Bulb nematode (*Ditylenchus* spp.) commonly attacks onion, garlic, potato, pea and carrot etc. The nematodes spread from one area to another mainly through infested planting materials, water drains from infested areas into irrigation system, soil that adheres to implements, tyres of motor vehicles and shoes of plantation workers. Management recommendation through bio-pesticides, cultural practices, enrichment of FYM, Neem cake and other organic amendments.

Keywords: fruits, vegetables, nematodes, symptoms, management

1. Introduction

Nematodes are microscopic roundworms live in soil, marine, freshwater. Plant parasitic nematodes cause economic damage to cultivated crops in the tropics and subtropics areas, estimated about 10 percent of world crop production is lost due to nematode [1]. More than 4100 species of plant-parasitic nematode of global food security [2] and damage caused by plant nematodes has been estimated at $US80 billion per year [3]. Presently 25 genera of plant parasitic nematodes, include species that are economic pests of crop plants. Ten most important nematode genera are significance at global level *viz., Meloidogyne, Pratylenchus, Heterodera, Ditylenchus, Globodera, Tylenchulus, Xiphinema, Radopholus, Rotylenchulus and Helicotylenchus* [4]. The root-knot nematode (*Meloidogyne* spp.) include over 100 species, with *Meloidogyne incognita, Meloidogyne javanica*, *Meloidogyne arenaria*, and Meloidogyne hapla, are the most destructive to agricultural crop [5]. Root knot nematodes develop primary feeding site of the giant cell [6]. Multi-nucleated giant cell -induced within the host cell in the absence of cytokinesis. Cysts nematodes (*Heterodera* and *Globodera* spp.) eggs in cyst body of the female releases larvae (J_2) infect the host and develop into adult stages within host tissue. Cyst nematodes enter root tips and induce specialized feeding structures syncytia. [7, 8]. Root-lesion nematodes (*Pratylenchus* spp.) are distributed worldwide and wide host range of plant species [9]. Lesion nematodes are migratory endoprasite, feeding mainly in the root cortex. Typically symptoms are lesion formation on roots and aboveground chlorosis of leaf [10]. *Radopholus similis (*burrowing nematode) is a migratory plant parasitic nematode causes severe economic losses in yields and quarantine plant pest worldwide [11]. considered the most important phytopathogenic nematode in banana-growing areas and also damages the crop banana, citrus, pepper, coffee [12].

Plant Parasitic nematodes are associated in agricultural crop in global food security. Agriculturally important root-knot nematodes and identified by Berkeley [13] (1855) who observed galls on cucumber roots. Plant-parasitic nematodes have a stylet, which is used for penetration of host plant tissue and release proteinaceous secretions from the glands to the host cell. These glandular secretions induce cellular metabolically active feeding cell [14]. Cellulose is the primary component of plant cell walls, cellulases (β-1,4-endoglucanases) are secreted to degrade the cell wall which allows nematode entry into host tissue. On the basis of their feeding habits, they are migratory ectoparasites, endoparasites, semi-endoparasitic. Ectoparasitic nematodes in the soil, feed at the root surface and Endoparasitic nematodes feed within the root. Endoparasitic nematodes are further divided into migratory and sedentary groups. Migratory endoparasitic nematodes include *Pratylenchus* spp. (lesion nematode), *Radopholus* spp. (burrowing nematodes) and

Hirschmanniella (rice root nematode). Nematodes associated with cultivated crops that are considered economically important. Viz., cyst nematodes (*Heterodera* spp.), lesion nematodes (*Pratylenchus* spp.); root knot nematodes (*Meloidogyne* spp.); and stem nematode (*Ditylenchus dipsaci)*. Plant parasitic nematodes feed underground plant tissues, such as roots, rhizomes, tubers, bulbs and symptoms appears on the aerial parts and often confused with those from abiotic stress, such as lack of nitrogen and water deficit. Some nematode species feed stems, leaves, flowers, fruits and seeds. *Some Plant parasitic nematodes are highly polyphagous in nature, such as Meloidogyne spp. and Pratylenchus spp.,* which can infect many species of plants [15].

The most economically important nematodes, the root-knot and cyst nematodes are wide range of species [16]. The potato (*Solanum tuberosum*) crops suffering root-knot nematodes (*Meloidogyne* spp.), stem nematode *Ditylenchus destructor* and cyst nematodes are cause losses in yields. Root-knot nematode (*Meloidogyne chitwoodii*) is considered the most important species [17]. *Globodera rostochiensis* and *Globodera pallida* originate from S. America are known pests [18]. In sweet potato (*Ipomoea* batatas L. Lam) *Ditylenchus destructor* is a major pest cause up to 100% yield losses in China [19, 20]. Lamberti [21] (1979) reported 50–60% losses in tomato and eggplant by RKN. Bhatti and Jain [22] estimated a loss of 46.2% in tomato due to *M. incognita.* Alam and Jairajpuri [23] estimated that nematodes are responsible for causing up to 70–90% yield losses in tomato and brinjal.

Pratylenchus coffeae in an area of about 1,000 ha in Karnataka was estimated to be Rs. 25 million is assessed in coffee. *G. rostochiensis* in Nilgiris, about 3000 ha area is infested with this nematode and total crop failure. *Meloidogyne* spp. attack more than 3000 crop plants, include vegetables, tuber crops, pulses, number of fruits, ornamental crops, tobacco etc. In India, the nematodes that cause most severe damage to horticultural crops *viz., Meloidogyne* and *Rotylenchulus reniformis* in vegetables, *Radopholus similis* in banana, black pepper and coconut (toppling disease of banana, slow wilt of pepper and coconut), *Pratylenchus coffeae* in coffee, *Tylenchulus semipenetrans* in citrus (Citrus decline/Slow decline of citrus). Root-knot nematode (*Meloidogyne* spp.) ranks first damage at global level and worldwide distribution, wide host range, destructive nature caused disease complexes. Plant parasitic nematodes often interact with fungal, bacterial and viral pathogens to cause disease complexes. Solanaceous vegetables yield losses by root-knot nematode have been assessed in various parts of the country. Developing countries suffer a crop loss of 14.6% compared to 8.8% in developed countries.

Fruits are the most important rich in Vitamins A and C and minerals like Calcium and Iron, low caloric values and low in fats. The plant parasitic nematodes are economic importance in fruit production. Fruit crops are perennial in nature, harbor and build-up of nematode population. Roots damaged by the nematodes lose efficiency in the utilization of available soil moisture and nutrients and easy prey to many fungi and bacteria which cause root decay. Symptoms of nematode attack often include reduced growth, chlorosis, wilting and death of plants. These resulted in reduced yields and poor fruit quality of fruits viz., citrus, banana, grapevine, pineapple, pomegranate and papaya. Nematode management is important for high yields and quality of fruits production. The integrated Nematode Management is population reduction of plant parasitic nematodes and development of resistant varieties of crops.

1.1 Root-knot nematodes (*Meloidogyne* spp.)

Root-knot nematodes (*Meloidogyne* spp.) are obligate parasites and are a major economic importance distributed world-wide. Root-knot nematodes (*Meloidogyne* spp.) attack in tropical and subtropical areas of crops by *M. incognita*, *M. javanica* and *M. arenaria* are known [24]. *Meloidogyne* females almost spherical in shape, deposit eggs in an egg sac, galling of the roots is common. Eggs are broken by temperature increasing after a cool period [25]. The *Meloidogyne* life cycle consists of six developmental stages; egg, four larval stages (juvenile) and adult, taking approximately 25 days for completion. Second larval stage (J2) able to penetrate and parasite the host cells [26, 27]. Lifecycle involves four developmental stages, larval stage (J_1) within the egg, larval stage (J_2) migratory, larval stage (J_3) sedentary, larval stage (J_4) sedentary and adult stage (sedentary). Females are characteristic "apple" shape Greek nomenclature *Meloidogyne*. Infective second-stage juveniles (J2s) are often attracted to root exudates and migrate to root tips behind the root cap at the elongation zone. The Meloidogyne genus well recognized species, *M. incognita, M. javanica, M. arenaria,* and *M. hapla* are the most important ones [28]. Root-knot nematodes are major pests of horticultural crops and infect more than 50 horticultural plant species. Meloidogyne is cosmopolitan genus, and distribued in temperate, subtropical, and tropical areas [29]. In warmer regions on volunteer grass hosts, more than one generation per season is possible [30]. Plant parasitic nematodes present some of the most difficult pest problems evaluated in our agricultural economy, because nematode damage is often overlooked due to mostly non specific symptoms. Mature females lay eggs in a protective gelatinous matrix which forms an egg mass. *Meloidogyne* species have a broad host range and, in general, hatching is dependent solely on suitable temperature and moisture conditions, with no stimulus from host plants being required. After embryogenesis, the infective second-stage juvenile (J_2), hatches from the egg. J_2 usually penetrate the roots directly behind the root cap, combination of physical damage through thrusting of the stylet and breakdown of the cell wall by cellulolytic and pectolytic enzymes. The J2 initiates a permanent feeding site, consists of several giant cells. These cells function as specialized sinks, supplying nutrients to J2. The head is embedded in the periphery of the vascular tissue. Symptoms in patches of poorly growing, yellowing plants few square metres to larger areas. Typical symptoms include stunted growth, wilting, leaf discoloration and deformation of the roots. The increased metabolic activity in giant cells mobilizes photosynthetic products from shoots to roots [31]. The damage by some *Meloidogyne* species has been given by Wesemael *et al.* [32]. Disease complexes with *Fusarium* wilt, *Rhizoctonia solani* and *Thielaviopsis basicola,* have been reported (Manzanilla-López and Starr, 2009). Many vegetable crops are susceptible to *Meloidogyne* spp., such as potato, tomato, carrot, lettuce, okra, cucumber and gourds. The root-knot damage in potato can be recognized by plants withering even in moist soil and yellowed leaves and stunting growth [33]. *Meloidogyne hapla* causes more root gall proliferation in potato.

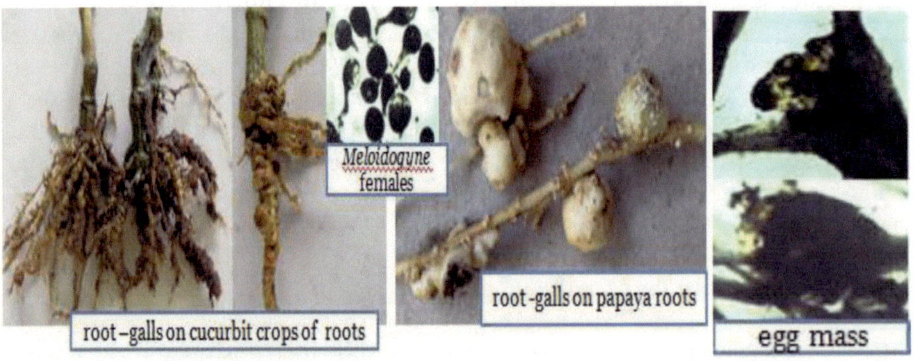

Figure 1.
Root-knot of infected roots, Meloidogyne females and egg mass.

Root-knot infection symptoms in ginger formation of gall characterized on rhizome showing brown to black lesions with accumulated fluid [34, 35]. In carrots, infection results in excessive root growth, forking formations and galls around the lateral roots and galls can be easily observed in okra and pumpkin roots [34, 36]. Among the fruits the root-knot nematodes hosts of banana (*Musa* spp.), grapevine (*Vitis vinifera*), and papaya (*Carica papaya*) **Figure 1**.

1.2 Cyst nematodes (*Heterodera* spp.)

Cyst nematodes (*Heterodera* spp.) attack wide range of pulses, grains, and vegetables including carrots, beans and peas cause huge economic losses for farmers. Vegetables *viz*., carrots, beans, peas, beetroot, cyst nematodes cause significant losses to agricultural crops. The beet cyst nematode (*Heterodera schachtii*) can cause yield loss in vegetable crops such as cabbages, Chinese cabbages, cauliflowers, Brussels sprouts, broccoli, turnip, radish beets and spinach, damaging root especially during summer. Nematodes feed on plant roots, reducing the ability of plants to take up nutrients and water. The above-ground symptoms look like nutrient deficiency, poor growth, stunting, yellowing and wilting. The sign of beet cyst nematode is the appearance of glistening white-yellow bodies about the size of a pin head attached to the fibrous roots. These mature and harden to produce a light-brown to reddish-brown cyst. Cyst nematodes (*Heterodera* spp.) are one of the most important groups of plant-parasitic nematodes worldwide. The most economic importance on cereals cyst nematode (*Heterodera avenae)*, detected in India [37, 38]. *Heterodera avenae* has polymorphous with many pathotypes [39, 40]. *H. avenae* is sexual dimorphism, female lemon-shaped and eggs are retained within body, after the female died, body wall hardens resistant brown cyst, which protects the eggs and juveniles. The eggs within the cyst remain viable for several years [30] (**Figure 2**).

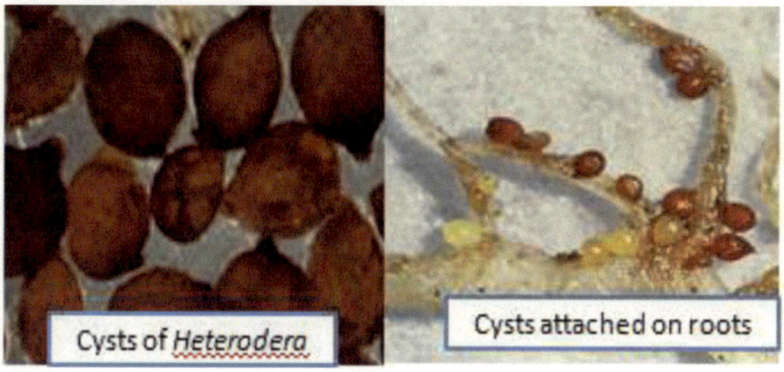

Figure 2.
Cyst of Heterodera sp. and cyst attached with the roots.

1.3 Lesion nematodes (*Pratylenchus* spp.)

The genus *Pratylenchus* is a large group of species affecting both monocots and dicots, polyphagous migratory endoparasites. Lesion nematodes (*Pratylenchus* spp.) are most common genus and have a very wide host range crops such as vegetables, and tree fruits. In fruit orchards, this nematode can be a major cause of orchard replant failures. It major problem causing damage in apple, peach, cherry, grape and potato. Eggs are lay inside root tissues and emerging juveniles enter in the roots and cause root injury, and allow other soil microorganisms to enter the root tissues and contribute to root rot and decay. Root lesion nematodes are migratory and capable of repeatedly entering and exiting from root tissue, several generations occur inside the roots. Nematodes cause small brown lesions on the white lateral roots and kill the fine feeder roots. Affected plants lose all feeder roots, its serious problem in orchard replant. The infected tree often exhibits stunting, chlorosis, and twig dieback with a decline in vigor, especially in peach and cherry orchards. The nematode invades the tissues of the plant root, migrating and feeding inside the root causing characteristic dark brown or black lesions on the root surface.

Pratylenchus pratensis was first described in 1880 [9]. Currently, the genus *Pratylenchus* is found worldwide, with more than 75 species described [41]. Lesion nematodes an important group of root migratory ecto-endoparasites in a variety of hosts. The damage caused by lesion nematodes in roots and tubers. *Pratylenchus* attack host tissues due to their free movement through the root and feeding in the plant cortex, resulting in dark spots or lesions. Members of the *Pratylenchus* genus may be sexual or asexual depending on the species. *Pratylenchus neglectus, P. thornei* and *P. brachyurus* reproduce by parthenogenesis, while *P. penetrans* must mate before producing fertile eggs. *Pratylenchus thornei* is considered the most economically important species. *Pratylenchus* females lay an average of one egg per day, the life cycles range from 45 to 65 days [42, 43]. All stages are capable of infecting and feeding from plant roots. Lesion nematode damage in plant roots is generally evidenced by necrosis and death of plants. The most economically important species in potato and tubers worldwide are *P. penetrans* and *P. scribneri* [34]. Root lesion nematodes get inside tubers using lenticels and surrounding tissue causing variable-sized circular lesions [44]. These lesions are usually superficial and decrease the marketable quality [9]. The symptoms of root lesion nematode infection in yam are more severe, and infection is characterized by a dehydrated, cracked skin, and softness of the tuber. *Pratylenchus* attack in broadleaf plants, especially lettuce, results in a reduction in growth, yellowing, and small head formation [36] **Figure 3**.

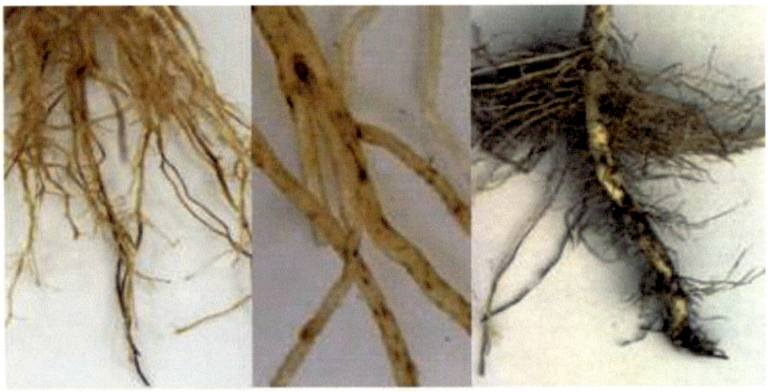

Figure 3.
Lesions on roots caused by Pratylenchus sp.

1.4 Burrowing nematodes (*Radopholus* spp.)

Radopholus genus is known as burrowing nematode, species are migratory endoparasitic. The *Radopholus* genus has a relatively wide host range and the most important species for horticulture crops. *Radopholus citri* and *Radopholus similis* infect citrus and banana respectively. Burrowing nematode causes damage worldwide and found especially in tropical and subtropical areas [45]. *Radopholus* genus similar behavior and life cycles as *Pratylenchus* spp. Burrowing nematode highly infective and complete their life cycles within the host root cortical cells, causing root cell death and necrosis. *Radopholus* reproduces sexually; 2 to 5 eggs are laid per day in the root tissues, and life cycle completed in 3 weeks under favorable conditions [46]. The most characteristic symptoms of infected plants are lesions, burrows in the roots and malnutrition [34]. The disease caused by *Radopholus similis* in banana is known as 'blackhead' due to feeding on the root cortex, cause internal necrosis, symptoms on above ground parts of banana are stunted growth, yellowing leaves, small bunches and uprooting [47]. *Radopholus similis* has already been recorded as the cause of 100% losses in Cavendish banana [48]. *Radopholus similis* is also a problem for ginger, and the symptoms are similar to those seen in *Meloidogyne* infection. *Radopholus citri* and *Radopholus similis* are serious problems for citrus, causing stunted growth and reducing fruit in terms of quantity and quality **Figure 4**.

Figure 4.
Infected banana plan and lesion on roots.

1.5 Citrus nematode *(Tylenchulus semipenetrans)*

The citrus nematode was first discovered in California, later described as a new species, *Tylenchulus semipenetrans*, and causal agent of slow decline in citrus Cobb [49]. Citrus nematode (*Tylenchulus* sp.) is a semi-endoparasitic nematode cause's slow decline in citrus crops, it's also problem in grapes and apple. Second stage juveniles (J_2) are infective; partly invade roots and exposed part of female body becoming enlarged on the surface of the roots. The highest numbers of nematodes are found in late spring and late autumn. Infected plants show aerial symptoms similar to nutrient deficiencies. Damage root system of fruit trees in all soil types and. Having high pH. *Tylenchulus semipenetrans* reported in every citrus growing areas of the world [50]. *Tylenchulus semipenetrans* disseminated in citrus growing areas by infected preparative plant material. Higher populations are found in orchards in sandy soils with high organic matter [51]. Mature females are attached to roots covered by soil particles stick to the gelatinous matrix. *Tylenchulus semipenetrans* is a dimorphic species at both the juvenile and adult stage. The female juveniles feeding on the epidermis and superficial layers of the cortical parenchyma of the roots. The immature female penetrates into the deep cortical layers. It becomes sedentary and establishes a permanent feeding site consisting of specialized cells (nurse cells). The posterior portion of body swells and protrudes from the root surface while its elongate neck and head remains embedded into the cortex. Mature females produce eggs that are embedded in a gelatinous matrix secreted through the excretory pore. The length of the female life cycle from egg to egg ranges from four to eight weeks [52]. *Tylenchulus semipenetrans* is a sexually reproducing species, occasionally reproduce by facultative parthenogenesis. Nematod has a restricted host range, includes citrus, trifoliate orange, grapevines, persimmon, and olive [53–55]. Currently, three biotypes are accepted, citrus, poncirus and mediterranean [55, 56]. Slow decline

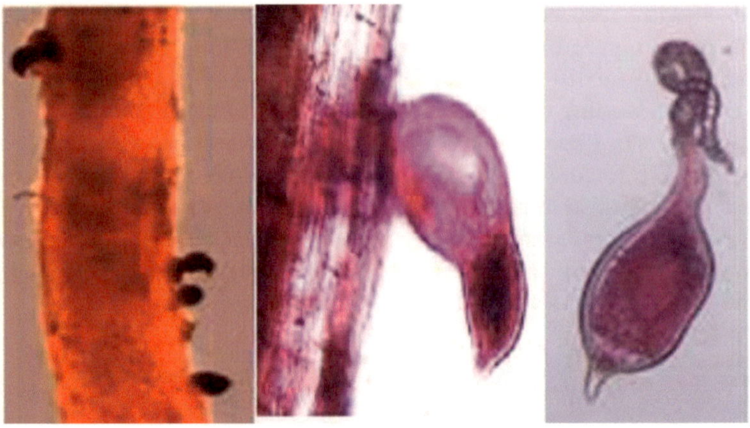

Figure 5.
Infected roots by Tylemchulus semipenetrans and female.

depend on age of trees and time of infection, reduced leaf and fruit size, most conspicuous symptoms of slow decline [50]. Yield losses due to *T. semipenetrans* range between 10–30% depending on the level of infection **Figure 5**.

1.6 Reniform nematode (*Rotylenchulus* sp.)

Reniform nematodes (*Rotylenchulus spp*) are a semi-endoparasitic species, females penetrate the root cortex and establish a permanent-feeding site in the stele region and become sedentary. The head region embedded in the root and tail region protrudes from the root surface and swells to form kidney-shaped. Genus *Rotylenchulus* have ten species; *Rotylenchulus reniformis* is the most economically important species [57] and called the reniform (kidney-shaped) nematode. *Rotylenchulus reniformis* is largely distributed in tropical, subtropical [58]. *Rotylenchulus reniformis,* has a wide host range on cultivated and noncultivated plants. First reported as a parasite of cotton in Georgia and tomato in Florida, and on cowpea roots in Hawaii [59]. Its associated with several kinds of tropical fruit trees [60–62]. Eggs hatch one to two weeks after laid. The second-stage juvenile (J_2) that emerges from the egg. *Rotylenchulus reniformis* is sexually dimorphic, males and females population are usually equal. Some populations of reniform nematodes reproduce parthenogenetically. Females produce eggs and deposited into a gelatinous matrix about 60 to 200 egg. The life cycle usually three weeks depends on soil temperature. It can survive two years in the absence of a host through anhydrobiosis [63] *R. reniformis* parasitizes a large number of plants and fruit trees. *R. reniformis* is a tropical nematode, thus soil temperature is not important factor [64]. Nematode causes root rotting and reduced uptake of water and soil nutrients. *R. reniformis* is pathogenic to sweet potato first time reported by Martin [65]. The reniform nematode causes root necrosis, dwarfing of plants, yellowing and wilting. The immature female formed C-shaped when killed by heat. The life cycle from egg to egg is from 22 to 29 days in susceptible host. Management recommendation by crop rotation with resistant plant species is recommended. These include mustard (*Brassica nigra*), oats, onion, sugarcane, and sun hemp [57, 66]. Sugarcane, Sorghum, maize and soybeans are recommended as rotation crops [67]. Managenet through crop rotation, non-host crops, resistant crops can be planted trap and antagonistic crops and use of organic mendments. Planting *Tagetes erecta* and *Crotolaria spectabilis* in nematode infested soil. *Paecilomyces lilacinus*, fungal egg parasite and effective against the reniform nematode **Figure 6**.

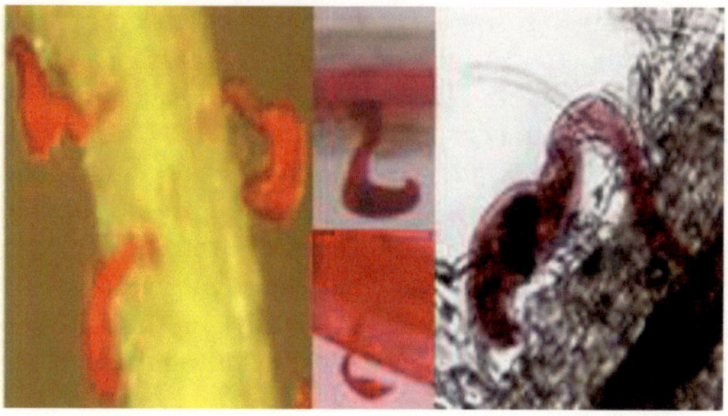

Figure 6.
Infected roots by Rotylenchulus sp.

1.7 Stem and bulb nematodes (*Ditylenchus* spp.)

Stem and bulb nematodes (*Ditylenchus* spp) are migratory endoparasitic nematodes that infect plant stems and leaves feeds upon parenchymatous tissue in stems and bulbs and continues in storage. Hosts crops are beans, onion, garlic, maize, oat, pea, potato, rye, strawberry, sugar beet, tobacco, alfalfa, faba, bersem, clover, and tulip. All stages of nematodes are infective. In adverse conditions species survive in dormant structure known as 'nematode wool', which is a bundle of juveniles [68, 69]. Fourth-stage juveniles aggregate just below the surface of infested tissue form "eelworm wool" and survive under dry conditions for several years. Runoff water is very important in the spread of stem nematodes within a field and to adjacent fields. In onion crop infected seedlings by *Ditylenchus dipsaci,* plant became twisted and deformed, leaves fall, bulbs become empty and roots yellow often death of plants. In garlic crops extensive longitudinal splitting of the cotyledons and leaves short and thick and often brown or yellowish spots, swelling above the bulb in the pseudo-stem [70]. Nematodes complete life cycle about 20 days in onion at 15° C. Infections are usually swollen, distorted stems, with reddish-brown to black lesions [71]. *D. dipsaci* and *D. destructor* causing dry rot in potato crops [69]. They enter potato tubers through the lenticels and multiply rapidly and invade the whole tuber

Figure 7.
Infected garlic and onion bulb by Ditylenchu sp.

and develop within tubers in storage. *Ditylenchus distructor* can also infect tulip and peanuts. However, horticultural losses caused by *Ditylenchus* have been most associated with *Allium* production [72]. *Ditylenchus dipsaci* and *Ditylenchus destructor* are important in commercial crops. *Ditylenchus dipsaci* is economic importance in temperate zones [30]. *Ditylenchus dispaci* is a migratory endoparasite that invades the foliage and the base of the stem. Characteristic symptoms of stem basal swellings, dwarfing and twisting of stalks and leaves, shortening of internodes and an abundance of axillary buds. *Ditylenchus dipsaci* have more than 10 biological races with limited host range. The only economic effective method is the use of host resistance [73]. Crop rotation with non-hosts crops including barley and wheat **Figure** 7.

2. Common practices in nematode management

Management practices should be effective, environmentally safe, and economical and must focus on reducing nematode populations to levels below the damage threshold. The common methods of nematodes management used resistant varieties, rotating of crops, soil amendments, soil solarization and applying pesticides. Soil solarization is very effective for control of many nematodes and soil-borne pathogens. Soil solarization of field to ensure adequate moisture, cover with plastic, to make it air tight, at least 45 days during June and July. Resistant plant cultivars is limited because few nematode very specific for specific resistance, correct identification of the nematode species and race before cultivar selected. Crop resistance cultivars with crop rotation is the best management practices.

In crop rotation, crops must be select carefully because some species of nematodes viz., root-knot, reniform, and burrowing are very wide host ranges. Crop rotation and cover cropping are often practices in integrated pest management to reduce plant-parasitic nematode incidence. Soil nematode effectively decreased by rotational cultivation of non-host cultivars of wide host range of *Meloidogyne spp*. [74]. Leguminous cover crops *Mucuna pruriens* L., and *Crotalaria spectabilis* showed multiple resistance to root-knot nematodes (*Meloidogyne arenaria, M. incognita*, *M. javanica*) [75]. Flooding and bare fallowing was shown to decrease nematode soil populations and increase strawberry yields [76]. Marigold, sudan grass and *Brassica* spp. can be used as green manure crops to control plant parasitic nematodes and boost free-living nematode populations in the soil. Glucosinilate or isocthiocyanate content in many *Brassica* species is known to control many plant parasitic nematodes. Cyst nematode can efficient manage through grass-free rotations using non-host crops. Clean fallow and deep summer plowing reduce the population density of the nematode. Cultivar resistance is considered one of the best methods for nematode control and has been found to be successful in several countries. Management of root lesion nematodes, the crop rotation is limited due to the polyphagous nature of the nematode. The role of crop rotation in controlling the lesion nematodes some field and laboratory work [77–79]. Cultural methods need to be integrated with other control measures. Mulching fields with polyethylene film for six to eight weeks suppressed *populations* by 50 percent [80]. Citrus nematodes can manage by use of resistant rootstocks and certified propagative citrus plants free from nematode for preventing the damage to citrus [81].

Green manuring as sudangrass and corn are excellent green manure crops that provide good nematode control. The organic agriculture for environmental welfare, biological controls are great interest for crop producers. The efficacy of nematophagous bacteria and fungi in the control of cyst and root-knot nematodes has been well-documented [82, 83]. Parasitic bacteria (*Pasteuria spp.*) have been reported to infect both plant-parasitic and free-living nematodes [84]. The application of

P. penetrans for nematode control *viz.*, seed, transplant, and post-plant treatments [85]. *Bacillus* spp. have shown great promise in nematode management. *B. cereus* strain S2 resulted in a mortality of 90.96% to *M. incognita* [86]. Inhibition of egg hatch and motility reported in *M. incognita*, [87]. Nematophagous fungi (*Pochonia chlamydosporia)* has potential as a biological control agent for *M. incognita* in vegetable crops. Nematophagous fungal products chitinases and their potential for the development of biopesticides. Plant extracts often contain a myriad of compounds which demonstrate nematode suppressive properties. Ethanolic extracts of *Azadirachta indica*, *Withania somnifera*, *Tagetes erecta*, and *Eucalyptus citriodora* were reported to nematicidial activity against *Meloidogyne incognita*, *Helicotylenchus multicinctus* and *Hoplolaimus* [88]. Organic matters contribute to biological activity in the soil and enhance the natural activity of organisms antagonistic to nematodes. Large populations of free-living nematodes can help control many different plant parasitic nematodes in the soil, so provide enough organic matter to increase free-living nematode populations. Natural biological control to incorporate soil amendments such as manure (poultry manure) and compost.

Most nematicides are highly toxic synthetic pesticides health risk. Limitations uses of chemical pesticides are alternative methods and great attention to nematode control. Chemical nematicides are often used in the management of root-knot nematodes, restrictions in some soil fumigants due to increased environmental toxicity expensive costs and risk to humans.

Nematode release β-1,4-endoglucanase and polygalacturonase during primary infection and feeding site and in plants growth proteins are secreted during processes to allow for cell enlargement [89]. Several root-knot resistance gene (*Mi-1*) identified in tomato [90]. In carrots, two root-knot nematode resistance genes *Mj1* [91] and *Mj2* [92] to *M. javanica*. Many specific genes involved in plant immune responses by root-knot nematodes [93]. Reactive oxygen species (ROS) accumulation is toxic to nematodes and lead to hypersensitive response during the response to root-knot nematode invasion [94]. Pathogenesis-related (PR) proteins have been identified based on their enzyme function [95]. The PR family are characterized b-1,3 glucanases, chitinases, proteinase inhibitors, defensins, ribonucleases and thionins. PR gene expression is often induced by ethylene, salicylic acid, jasmonic acid, xylanase, and systemin signaling pathways. PR transcripts accumulate in high concentrations with the long distance immune response termed systemic acquired resistance (SAR) [96]. The roles of plant developmental hormones, ethylene, jasmonic acid and salicylic acid boost up plant immunity [97]. Jasmonic acid (JA) and ethylene (ET) signaling pathways work synergistically while the salicylic acid (SA) pathway is antagonistic to JA/ET pathways [98]. Exogenous ethyhlene (ethephon) and jasmonic acid (methyl jasmonate) application triggered the induction of PR proteins and the activation of systemic defense against root-knot nematodes on rice [99]. The role of jasmonic acid in activation of systemically induced resistance, exogenous application of jasmonic acid and arachidonic acid, decreased galling on tomato roots [100]. The role of salicylic acid in host resistance against root-knot nematodes. Pathogenesis-related protein expression of salicylic acid-dependent systemic required resistance in tomatoes root-knot nematode [96]. Expression of a *NahG* which encodes for an enzyme that degrades salicylic acid to catechol, reduced *Mi-1* gene in root-knot. Nematode-resistant genes found in gene pools of a variety of plant species have been introgressed into the genomes of economically important crops through transgenic technologies [101, 102].

Management of nematodes is an integrated method of pest management system. Because of most commonly practiced methods including crop rotation, developing resistant and tolerant cultivars, using chemicals and cultural practices [24, 73]. Effective management practices are required accurate diagnosis, and proper effective management techniques.

3. Conclusion

There are several genera and species of nematodes that are of economic importance. Correct nematode diagnosis can developing management program. The nematodes must be eliminate minimize the damage to determine the appropriate method. Commonly practiced methods are including crop rotation, resistant and tolerant cultivars, cultural practices and chemicals. The ability to reduce yield losses caused by nematodes is need to understanding about pathogen biology and the application of appropriate control measures. Use of chemicals is impractical commercial and cultural methods fail to complete control. Breeding for resistance and tolerance is the major strategy for long-term and environmentally sound control. It is necessary to research particularly nematodes race and pathotype, and a great need for global collaborative research to control of these important pathogens.

Author details

Amar Bahadur
Plant Pathology, College of Agriculture, Tripura, Agartala, Tripura, India

*Address all correspondence to: amarpatel44@rediffmail.com; agcollege07@gmail.com

References

[1] Whitehead, A.G (1998) *Plant nematode control*. Wallingford, UK, CAB.

[2] Decraemer W, Hunt D (2006) Structure and classification. In: Perry R, Moens M, editors. Plant Nematology. Oxfordshire: CAB International. pp. 3-32.

[3] Nicol J., et al. "Current nematode threats to world agriculture". In: Jones J, Gheysen G, Fenoll C, editors. (2011). Genomics and Molecular Genetics of Plant-Nematode Interactions. Berlin: Springer Science Business Media (2011): 21-43.

[4] Jones J, Haegeman A, Danchin E, Gaur H, Helder J, Jones M, Kikuchi T, Manzanilla-Lopez R, Palomares-Rius J, Wesemael W, Perry R (2013) Top 10 plant-parasitic nematodes in molecular plant pathology. Molecular Plant Pathology.14: 946-961.

[5] Postnikova O, Hult M, Shao J, Skantar A, Nemchinov L (2015) transcriptome analysis of resistant and susceptible alfalfa cultivars infected with root-knot nematode Meloidogyne incognita. PloS One. 2015; DOI: 10.1371/journal.pone.0118269

[6] Davis E, Hussey R, Baum T, Bakker J, Schots A, Rosso M, Abad P (2000). Nematode parasitism genes. Annual review of phytopathology. (38): 365-396.

[7] Hewezi T, Baum T (2013) Manipulation of plant cells by cyst and root-knot nematode effectors. Molecular plant-microbe Interactions.26:9-16. DOI: 10.1094/MPMI-05-12-0106-FI

[8] Gheysen G, Fenoll C (2002) Gene expression in nematode feeding sites. Annual Review of Phytopathology.40: 191-219.

[9] Davis E and MacGuidwin A (2000) Lesion nematode disease. The Plant Health Instructor. [Internet]. 2000. DOI: 10.1094/PHI-I-2000-1030-02

[10] Jones M, Fosu-Nyarko J (2014) Molecular biology of root lesion nematodes (*Pratylenchus* spp.) and their interaction with host plants. Annals of Applied Biology. 164: 163-181.

[11] Smith I, Charles L (1998) Distribution maps of quarantine pests for Europe. Wallingford: CABI International; 1998. pp. 1-78

[12] Sarah J, Gowen S, De Waele D, Tessera M, Quimio A (1999) Nematode pathogens. In: Jones D, editors. Diseases of Banana, Abacá and Ensete. Wallingford: CABI Publishing; pp. 295-303

[13] Berkeley M (1855) Vibrio forming cysts on the roots of cucumbers. Gardeners' Chronicle. 7: 220

[14] Davis EL, Hussey RS, Mitchum MG and Baum TJ (2008) Parasitism proteins in nematodeplant interactions. Current Opinion in Plant Biology.11: 360-366

[15] Palomares-Rius JE, Hedley PE, Cock PJ, Morris JA, Jones JT, et al. (2012) Comparison of transcript profiles in different life stages of the nematode Globodera pallida under different host potato genotypes. Mol Plant Pathol 13: 1120-1134.

[16] Jacob, J. and Mitreva, M. (2011) Transcriptomes of plant parasitic nematodes. In: Genomics and Molecular Genetics of Plant–Nematode Interactions (Jones, J.T., Gheysen, G. and Fenoll, C., eds), pp. 119-138. Heidelberg: Springer.

[17] Santo G, O'Bannon J, Finley A, Golden A (1980) Occurrence and host range of a new root-knot nematode (*Meloidogyne chitwoodi*) in the Pacific northwest. Plant Disease. 64: 951-952.

[18] Turner S, Evans K (1998) The origins, global distribution and biology of potato cyst nematodes (*Globodera rostochiensis* (Woll.) and *G. pallida* Stone). In:

Marks R, Brodie B. editors. Potato Cyst Nematodes: Biology, Distribution, and Control. Cambridge: University Press; pp. 7-26

[19] Xu Z, Zhao Y, Yang D, Sun H, Zhang C, Xie Y (2015) Attractant and repellent effects of sweet potato root exudates on the potato rot nematode, Ditylenchus destructor. Nematology. 17: 117-124.

[20] Zhang S, Zhang S, Wang H, Chen Y (2006) Characteristics of sweetpotato stem nematode in China. Acta Phytopathologica Sinica. 36: 22-27

[21] Lamberti F (1979) Economic importance of *Meloidogyn*e spp., in sub-tropical and Mediterranean climate. In: Root knot nematode (*Meloidogyne* species). Systemics, biology and control. (Eds.) Lamberti F and Taylor CE, academic press, London, UK. 341-357.

[22] Bhatti DS, Jain RK (1977) Estimation of losses in okra, tomato and brinjal yield due to *Meloidogyne incognita.* Indian Journal of Nematology,7:37-41.

[23] Alam MM, Jairajpuri MS (1990) Nematode Control Strategies. In: Nematode Bio-Control (Aspects and Prospects). M.S. Jairajpuri, M.M. Alam and I. Ahmad (Eds.). CBS Pub. and Dist. Delhi, India. 5-15.

[24] Swarup, G. and Sosa-Moss, C (1990) Nematode parasites of cereals. *In* M. Luc, R.A. Sikora & J. Bridge, eds. *Plant Parasitic Nematodes in Subtropical and Tropical Agriculture*, p. 109-136. Wallingford, UK, CAB International.

[25] Antoniou, M (1989) Arrested development in plant parasitic nematodes, Helminth. Abstr. Ser. B, 58: 1-9.

[26] Ibrahim IKA, Massoud SI (1974) Development and pathogenesis of a root-knot nematode, Meloidogyne javanica. Proc Helm Soc Wash 41: 68-72.

[27] Maleita C, Curtis R, Abrantes I (2012) Thermal requirements for the embryonic development and life cycle of *Meloidogyne hispanica*. Plant Pathol 61: 1002-1010.

[28] Elling AA (2013) Major emerging problems with minor Meloidogyne species. Phytopathol 103: 1092-1102.

[29] Xalxo PC, Karkun D, Poddar AN (2013) Rhizospheric fungal associations of root knot nematode infested cucurbits: In vitro assessment of their Nematicidal potential. Res J Microbiol 8: 81-91.

[30] Kort, J (1972) Nematode diseases of cereals of temperate climates. *In* J.M. Webster, ed. *Economic nematology*, p. 97-126. New York, NY, USA, Academic.

[31] Hofmann, J. and Grundler, F.M.W. (2007) How do nematodes get their sweets? Solute supply to sedentary plant-parasitic nematodes. Nematology, 9, 451-458

[32] Wesemael, W.M.L., Viaene, N. and Moens, M (2011) Root-knot nematodes (*Meloidogyne* spp.) in Europe. Nematology, 13: 3-16

[33] Wesemael WML, Moens M, Viaene N, Taning LM (2014) Life cycle and damage of the root-knot nematode Meloidogyne minor on potato, Solanum tuberosum. Nematol 16: 185-192

[34] Bridge J, Starr JL (2007) Plant nematodes of agricultural importance: A colour handbook (1st Edn) CRC press, USA

[35] Okorocha EOA, Ogbuji RO, Ijeoma OF, Okorocha CG (2014) Relationship between root-knot nematode Meloidogyne javanica inoculum densities and ginger (*Zingiber officinale* roscoe). Sch Acad J Biosci 2: 809-812.

[36] Pinheiro JB, Amaro GB, Pereira RB (2010) Occurrence and control of

nematodes in leafy vegetables (Ocorrência e controle de nematoides em hortaliças folhosas) Embrapa, Brazil.

[37] Sharma, S.B. & Swarup, G (1984) Cyst forming nematodes of India. New Delhi. *Ind. Cosmo Publ.*, 1: 150.

[38] Sikora, R.A (1988) Plant parasitic nematodes of wheat and barley in temperature and temperate semi-arid regions - a comparative analysis. *In* M.C. Saxena, R.A. Sikora & J.P. Srivastava, eds. *Nematodes Parasitic to Cereals and Legumes in Temperate Semi-Arid Regions*, p. 46-48. Aleppo, Syria, ICARDA.

[39] Andersen , S. and Andersen , K (1982) Suggestions for determination and terminology of pathotypes and genes for vresistance in ncyst-forming nematodes especially *Heterodera avenae* EPPOBull., 12: 379-386.

[40] Cook, R and Rivoal , R (1998) Genetic of resistance and parasitism. *In* SB. Sharma, ed. The cyst nematodes. Dordrecht, Netherlands, Kluwer Academic.

[41] Araya TZ, Padilla WP, Archidona-Yuste A, Cantalapiedra-Navarrete C, Liébanas G (2016) Root-lesion nematodes of the genus Pratylenchus (Nematoda: Pratylenchidae) from Costa Rica with molecular identification of P. gutierrezi and P. panamaensis topotypes. Eur J Plant Pathol 145: 973-998.

[42] Ryss AY (2002) Genus Pratylenchus Filipjev : Multientry and monoentry keys and diagnostic relationships (Nematoda: Tylenchida: Pratylenchidae). Zoosystematica Ross 10: 241-255.

[43] Collins S, Wilkinson C (2015) Pratylenchus penetrans: A horticulturally significant root lesion nematode.

[44] Pinheiro JB, Amaro GB, Pereira RB (2011) Nematoides in Capsicum chili peppers [Nematoides em pimentas do gênero Capsicum] Embrapa Hortaliças, Brazil.

[45] Luc M, Bridge J, Sikora RA (2005) Reflections on nematology in subtropical and tropical agriculture. In: Luc M, Sikora RA, Bridge J (eds) Plant parasitic nematodes in subtropical and tropical agriculture. CABI, Wallingford, UK 1-12.

[46] Brooks FE (2008) Burrowing nematode diseases. Plant Heal Instr 42: 142-145

[47] Dias-Arieira CR, Molina R de O, Costa AT (2008) Disease-Causing Nematodes In Fruit Trees (Nematóides Causadores de Doenças em Frutíferas). Agro Ambient 2: doi: 10.18227/1982-8470ragro.v2i1.230.

[48] Pinheiro JB, Silva GO, Pereira RB (2015) Nematoids in potato culture [Nematoides na Cultura da Batata] Embrapa Hortaliças, Brazil.

[49] Cobb NA (1913) Notes on Mononchus and Tyenchus, Journals of the Washington Academy of Science, 3: 287-288

[50] Duncan LW (2005) Nematode parasites of citrus. pp. 437-466. *In* Plant Parasitic Nematodes in Subtropical and Tropical Agriculture (Luc M, Sikora RA, Bridge J, eds). CAB International, Wallingford, UK.

[51] Timmer LW, Garnsey SM, Broadbent P (2003) Diseases of citrus. Pp. 163-196. *In* diseases of tropical fruit crops. CAB international, Wallingford, UK.

[52] van Gundy SD (1958) The life history of the citrus nematode, *Tylenchulus semipenetrans* Cobb. Nematologica, 3: 283-294.

[53] Thorne G (1961) Principles of nematology. McGraw-hill book company, Inc. New York, NY

[54] Baines RC, Miyakawa T, Cameron JW, Small RH (1969)

Infectivity of two biotypes of the citrus nematode on citrus and on some other hosts. Journal of Nematology 1: 150-159.

[55] Inserra RN, Duncan LW, O'Bannon JH, Fuller SA (1994) Citrus nematode biotypes and resistant citrus rootstocks in Florida. Nematology circular No. 205. Florida Department of Agriculture and Consumer Services. Division of plant industry.

[56] Inserra RN, Vovias N., O'Bannon JH (1980) A classification of *Tylenchulus semipenetrans* biotypes. J. of Nematology 12: 283-287

[57] Robinson AF, Inserra RN, Caswell-Chen EP, Vovlas N, Troccoli A (1997) Rotylenchulus species: Identification, distribution, host ranges, and crop plant resistance. Nematropica 27: 127-180.

[58] Ayala A, Ramirez CT. 1964. Host-range, distribution, and bibliography of the reniform nematode, Rotylenchulus reniformis, with special reference to Puerto Rico. Journal of Agriculture of University of Puerto Rico 48: 140-160.

[59] Linford MB, Oliveira JM. 1940. Rotylenchulus reniformis, nov. gen. n. Sp., a nematode parasite of roots. Proceeding of the Helminthological Society of Washington 7: 35-42.

[60] McSorley R (1980) Nematodes associated with sweetpotato and edible aroids in southern Florida. Proceedings of Florida State Horticultural Society 93: 283-285.

[61] McSorley R, Campbell CW, Parrado JL (1982). Nematodes associated with tropical and subtropical fruit trees in South Florida. Proceedings of Florida State Horticultural Society 95: 132-135.

[62] McSorley R, Parrado JL, Conover RA (1983) Population buildup and effects of the reniform nematode on papaya in southern Florida. Proceedings of Florida State Horticultural Society 96: 198- 200.

[63] Radewald JD, Takeshita G (1964) Desiccation studies on five species of plant-parasitic nematodes of Hawaii. Phytopathology 54: 903-904.

[64] Heald, C.M. and R.N. Inserra (1988) Effect of temperature on infection and survival of *Rotylenchulus reniformis*. J. Nematology, 20(3): 356-361.

[65] Martin (1960) The reniform nematode may be a serious pest of the sweetpotato. Plt. Dis. Rept. 44: 216.

[66] Caswell EP, deFrank J, Apt WJ, Tang C-S (1991) Influence of nonhost plants on population decline of Rotylenchulus reniformis. Journal of Nematology 23: 91-98.

[67] Starr JL, Page SL (1990) Nematode parasites of cotton and other tropical fiber crops. pp. 539-556. In: Plant parasitic nematodes in subtropical and tropical agriculture. Luc M, Sikora RA, Bridge J (eds). CAB International, Oxon, UK.

[68] Bosher JE (1960) Longevity in vitro of Ditylenchus dipsaci (Kiihn) Filipjev from narcissus. The helminthological Society of Washington. Washington, USA 1960: 127-128.

[69] Mwaura P, Niere B, Vidal S (2015) Resistance and tolerance of potato varieties to potato rot nematode (Ditylenchus destructor) and stem nematode (Ditylenchus dipsaci) Ann Appl Biol 166: 257-270.

[70] Pinheiro JB, Ferreira AD, Carvalho D, Pereira RB (2014) Nematodes in garlic and onion culture (Nematoides na cultura do alho e cebola) Embrapa Hortaliças, Brazil.

[71] Zhang SL, Liu GK, Janssen T, Zhang SS, Xiao S, et al. (2014) A new stem nematode associated with peanut pod rot in China: Morphological and

molecular characterization of Ditylenchus arachis n. sp. (Nematoda: Anguinidae). Plant Pathol 63: 1193-1206.

[72] Fan W, Wei Z, Zhang M, Ma P, Liu G, et al. (2015) Resistance to Ditylenchus destructor Infection in Sweet Potato by the Expression of Small Interfering RNAs Targeting unc-15 , a Movement-Related Gene. Phytopathology 105: 1458-1465.

[73] Rivoal, R. & Cook, R (1993) Nematode pests of cereals. In *Plant parasitic nematodes in temperate agriculture*, p. 259-303. Wallingford, UK, CAB International.

[74] Sikora R, Fernandez E. Nematode parasites of vegetables. In: Luc M, Sikora R, Bridge J (2005), Plant parasitic nematodes in subtropical and tropical. Wallingford: CABI;. pp. 319-392.

[75] Osei K, Gowen S, Pembroke B, Brandenburg R, Jordan D (2010) Potential of leguminous cover crops in management of a mixed population of root-knot nematodes (*Meloidogyne* spp.). Journal of Nematology. 42: 173-178

[76] Chen P, Tsay T (2006) Effect of crop rotation on Meloidogyne spp. and Pratylenchus spp. populations in strawberry fields in Taiwan. Journal of Nematology.38: 339-344.

[77] Clewett, T.G., Thompson, J.P. & Fiske, M.L (1993) crop rotation to control *Pratylenchus thornei*. *In* V.a. Vanstone, S.P. Taylor & J.M. Nicol, eds. *Proc. 9th biennial Australian plant pathology Conf.* Pratylenchus *workshop*, Adelaide, Australia.

[78] Nicol, J.M (1996) The distribution, pathogenicity and population dynamics of *Pratylencus thornei* (Sher and Allen, 1954) on wheat in south Australia. Ph.D. thesis. Adelaide, Australia, The University of Adelaide.

[79] Hollaway, C.J., Taylor, S.P., Eastwood, R.F. & Hunt, C.H (2000) Effect of field crops on density of *Pratylenchus* in south-eastern Australia. Part 2: *P. thornei*. *J. Nemat.*, 32(4): 600-608.

[80] Di Vito, M., Greco, N. & Saxena, M.C (1991) Effectiveness of soil solarization for control of *Heterodera ciceri* and *Pratylenchus thornei* on chickpeas in Syria. *Nemat. Med.*, 19: 109-11.\

[81] Kaplan, D. T., and J. H. O'Bannon. 1981. Evaluation and nature of citrus nematode resistance in Swingle citrumelo. Proceedings of the Florida State Horticultural Society 94: 33-36

[82] Stirling G (1991) Biological control of plant-parasitic nematodes. Wallingford: CAB international;. 282 p

[83] Meyer S (2003) United States department of agriculture – Agricultural Research Service research programs on microbes for management of plant-parasitic nematodes. Pest Management Science.;59: 665-670

[84] Chen Z, Dickson D (1998) Review of Pasteuria penetrans: Biology, ecology, and biological control potential. Journal of Nematology. 30:313-340.

[85] Kokalis-Burelle N (2015) Pasteuria penetrans for control of Meloidogyne incognita on tomato and cucumber and M. arenaria on snapdragon. Journal of Nematology. 47:207-213

[86] Gao H, Qi G, Yin R, Zhang H, Li C, Zhao X (2016) Bacillus cereus strain S2 shows high nematicidal activity against Meloidogyne incognita by producing sphingosine. Scientific Reports. 6: 28756.

[87] Xiong J, Zhou Q, Luo H, Xia L, Li L, Sun M, Yu Z (2015) Systemic nematicidal activity and biocontrol efficacy of Bacillus firmus against the root-knot nematode Meloidogyne incognita.World Journal of Microbiology and Biotechnology. 31: 661-667.

[88] Khan A, Sayed M,. Shaukat S, Handoo Z (2008) Efficacy of four plant extracts on nematodes associated with papaya in Sindh, Pakistan. Nematologia Mediterranea, 36: 93-98

[89] Bashline L, Lei L, Li S, Gu Y (2014) Cell wall, cytoskeleton, and cell expansion in higher plants. Molecular Plant. 7: 586-600.

[90] Milligan S, Bodeau J, Yaghoobi J, Kaloshian I, Zabel P, Williamson V (1998) The root-knot nematode resistance gene mi from tomato is a member of the leucne zipper, nucleotide binding, leucine-rich repeat family of plant genes. Plant Cell. 10:1307-1319

[91] Ali A, Matthews W, Cavagnaro P, Iorizzo M, Roberts P, Simon P (2014) Inheritance and mapping of Mj-2, a new source of root-knot nematode (Meloidogyne javanica) resistance in carrot. Journal of Heredity. 105: 288-291

[92] Simon P, Matthews W, Roberts P (2000) Evidence for simply inherited dominant resistance to Meloidogyne javanica in carrot. Theoretical and Applied Genetics. 100: 735-742.

[93] Oliveira J, Andrade N, Martins-Miranda A, Soares A, Gondim D, Araujo-Filho J, FreireFilho F, Vasconcelos I (2012) Differential expression of antioxidant enzymes and PR-proteins in compatible and incompatible interactions of cowpea (*Vigna unguiculate*) and the rootknot nematode Meloidogyne incognita. Plant Physiology and Biochemistry. 51: 145-152.

[94] Kawano T (2003) Roles of the reactive oxygen species-generating peroxidase reactions in plant defense and growth induction. Plant Cell Reports. 21:829-837.

[95] van Loon L, van Strien E (1999) The families of pathogenesis-related proteins, their activities, and comparative analysis of PR-1 type proteins. Physiological and Molecular Plant Pathology. 55: 85-97.

[96] Molinari S, Fanelli E, Leonetti P (2014) Expression of tomato salicylic acid (SA)-responsive pathogenesis-related genes in Mi-1-mediated and SA-induced resistance to root-knot nematodes. Molecular Plant Pathology. 152: 55-64.

[97] Loake G, Grant M (2007) Salicylic acid in plant defence-the players and protagonists. Current Opinion in Plant Biology. 10: 466-472.

[98] Glazebrook J (2005) Contrasting mechanisms of defense against biotrophic and necrotrophic pathogens. Annual Review of Phytopathology. 43:205-227.

[99] Nahar K, Kyndt T, De Vleesscchauwer D, Hofte M, Gheysen G. (2011) The jasmonate pathway is a key player in systemically induced defense against root-knot nematodes in rice. Plant Physiology. 157:305-316.

[100] Vasyukova N, Zinovieva S, Udalova Z, Gerasimova N, Ozeretskovskaya O, Sonin M (2009) Jasmonic acid and tomato resistance to the root-knot nematode Meloidogyne incognita. Doklady Biological Sciences.;428: 448-450.

[101] Vishnudasan D, Tripathi M, Rao U, Khurana P (2005) Assessment of nematode resistance in wheat transgenic plants expressing potato proteinase inhibitor (PIN2) gene. Transgenic Research.14:665-675.

[102] Matthews B, Beard H, Brewer E, Kabir S, MacDonald M, Youssef R (2014) Arabidopsis genes, AtNPR1, AtTGA2 and AtPR-5, confer partial resistance to soybean cyst nematode (*Heterodera glycines*) when over expressed in transgenic soybean roots. BMC Plant Biology. 14:1-19.

Chapter 2

The Emerging Nematode Problems in Horticultural Crops and Their Management

Saroj Yadav and Jaydeep A. Patil

Abstract

Plant-parasitic nematodes (PPNs) are responsible for significant monetary losses to horticultural crops. They are unseen foes of crops and devitalize plants by causing injury to plant roots or aboveground parts. From last few decades, increased attention has been paid to nematode problems in horticultural crops in open as well as under protected cultivation. PPNs are obligate parasites, mostly have wide host range and are widespread pathogens of horticultural crops. The dimension of damage is density dependent and their management options vary with type of crop, nematode species and other factors. Recent approaches to combat losses caused by nematodes are the use of nematicides, cultural practices and resistant cultivars that may be used singly or in an integrated manner. This book chapter gives an overview of the emerging nematode problems in horticultural crops and their management strategies.

Keywords: nematode, horticultural crops, disease complex, management

1. Introduction

Horticulture has emerged as an important and viable diversification option in agriculture which transformed the subsistence farming system into a high value commercial enterprise [1]. It has led to the revolutionary changes in the socio-economic status of farmers in various parts of the country. Due to the rich diversity of agro-climatic conditions prevailing in the country, India is regarded as a horticultural paradise. Estimated 234.2 tonnes of horticultural produce are being produced from about 20.66 million hectares [2]. The horticulture has proven its credibility as a potential sector to enhance agricultural production, improve household nutritional security and income generation through diversification and employment, value addition and export [3]. Innovative technologies and their execution coupled with supporting infrastructure has brought India at the doorstep of golden revolution, which has enabled India the global leader in production of many horticultural crops [4]. In spite of the enormous success achieved in the horticulture sector, several constraints still exist. Besides new emerging challenges, poor productivity per unit area continues to be a concern in most of the horticultural crops with climate change impacting productivity further. Phytonematodes are among the important biotic stresses in the horticultural sector cause severe losses in these crops (**Table 1**). The introduction of new species, new

S. No.	crops	Nematode affecting the crop	Per cent yield losses	Monetary losses (Rs. in millon)
1.	Banana	*Meloidogyne incognita*	15	9710.46
2.	Guava	*Meloidogyne* spp.	28	2350.88
3.	Citrus	*Tylenchulus semipenetrans*	27	9828.22
4.	Pomegranate	*Meloidogyne* spp.	23	3023.44
5.	Brinjal	*M. incognita*	21	3499.12
6.	Chili	*M. incognita*	15	744.90
7.	Cucumber	*Meloidogyne* spp.	12	110.46
8.	Okra	*Meloidogyne* spp.	19.5	2480.86
9.	Tomato	*Meloidogyne* spp.	23	6035.20
10.	Capsicum	*Meloidogyne* spp.	10	52.92

Kumar et al. [5].

Table 1.
The losses caused by plant parasitic nematodes to economically important crops in India.

mutant of the species and resistance among the present species to various environmental conditions are emerging threats to all horticultural crops.

2. Emerging nematode problems

2.1 Nematode as emerging threat to protected cultivation

In the present era protected cultivation is an emerging technology adopted by farmers on a large scale, for growing seasonal as well as off-season crops. Under protected conditions farmers' taken several high economic value crops such as vegetables and ornamentals. Growers can noticeably increase their income by cultivation of vegetables in off-season as the vegetables produced during their normal season generally do not fetch good returns due to glut in the markets. Polyhouse crops attacked by a number of insects, pests and diseases including plant parasitic nematodes, they are interfering with the successful cultivation of vegetables and ornamentals under protected structures. Favorable conditions such as moisture, temperature and continuous availability of host in polyhouses favor the multiplication of phyto-nematodes. Among plant parasitic nematodes, root-knot nematode, *Meloidogyne* spp. are the most overwhelming pest in protected structures. Root-knot nematodes have wide host range attacking on almost all the crops grown under polyhouses and causes significant damage. Many a times complete crop failure due to association of nematode and soil borne fungi (disease-complexes) have been reported (**Figure 1**). In Haryana, 63.15% polyhouses were found infested with root-knot nematodes [6].

Capsicum (bell pepper), cucumber, chillies, tomato, muskmelon, gerkins and ornamentals like chrysanthemum are very common crops in protected cultivation. These crops are grown throughout India and are seriously infested with *Meloidogyne incognita*, *Meloidogyne javanica* (Root-knot nematodes) and *Rotylenchus reniformis* (Reniform nematode). Other nematodes such as lesion nematode (*Pratylenchus* spp.), foliar nematode (*Aphelenchoides* spp.) and burrowing nematode (*Radopholus similis*) which may develop in vegetable and ornamental crops grown under polyhouses. The overall annual yield losses due to nematodes goes up to 60% under

Figure 1.
Nematode infestation under polyhouses a. root-knot nematode infested root B. crop failure (source: Original photos).

protected structures and also reduced the quality of the produce of crops in the market due to unthrifty growth.

2.2 The emerging threat of nematodes in orchards and plantation crops

Nematodes are serious menace in fruit orchards and plantation crops in the country. Recently, the threat has been increased due to the introduction of new nematode species such as *Meloidogyne enterolobii* is an emerging threat to guava [7]. Growers of fruits such as guava, pomegranate, citrus, banana are facing the nematode problems in their orchards such as root-knot nematodes, citrus nematode, reniform nematode, burrowing nematode and lesion nematode etc. Farmers are not aware to this menace due to hidden nature and non-diagnostic symptoms which is of major concern in their multiplications in the orchards. Recently, pomegranate and guava growers are encountering the problems of yellowing of leaves, stunting and less productivity of trees (**Figure 2**). Such trees were found to be severely infested with root-knot nematodes on the basis of soil and root samples. Crop losses caused by phyto nematodes to fruit crops are very high, with an average annual yield losses estimated at about 20–40% worldwide [5]. The losses have increased tremendously when two or more species occurs simultaneously or mainly with secondary pathogens like fungi and bacteria forming disease-complexes.

2.3 Potato cyst nematodes, *Globodera* species on potato

Outbreaks of potato cyst nematodes are reported in most of the potato growing temperate areas of the world and declared as a quarantine pest throughout the world. In India, it was reported from the Nilgiri hills during 1961and triggered the implementation of domestic quarantine in 1971 under the Destructive Insect Pest Act 1914. Under this act movement of potato for seed purposes was restricted realizing the potential threat to potato production and trade [8, 9]. However, recent reports on the presence of potato cyst nematode in some of the potato growing areas in our country have implications of the movement and production of potato. Recently it has been reported from Himachal Pradesh, Uttarakhand and Jammu and Kashmir [10].

Figure 2.
Guava orchard infested with root knot nematode (source: Original photos).

2.4 The emerging nematode problem in mushroom cultivation

Several biotic factors such as fungi, bacteria, nematodes, insects and mites have challenged mushroom cultivation. Among these, mushroom nematodes are the most dreaded ones. Once they get entry into the mushroom houses, it becomes almost impossible to get rid of these worms. Their multiplication is very fast because of very short life cycle (8–10 days). They attack on a variety of fungi, including plant pathogenic ones. They are stylet bearing like other plant parasitic nematodes and have the same feeding mechanism. Nematodes affected mushroom hyphae produced less yield by sucking cell sap. Their presence in mushroom beds is associated with severe losses, often leading to crop failure [11]. Unlike the management practices commonly used for plant parasitic nematodes like crop rotation, summer plowing, use of chemicals' etc., are not applicable here due to the different nature of the crop.

2.5 Nematodes in flowers

Apart from root-knot nematodes and lesion nematodes on various crops, *Aphelenchoides besseyi* has raised as new peril in floriculture. Interestingly, in recent years, as a single flower, tuberose, *Polianthes tuberosa* L. occupies the lion share in the global publications on the nematode problems in flowers. Phenological factors influenced nematode population towards consistent variation with several ups and downs synchronizing with that of the flower production. Heavy infestation at early stages of the plants resulted stunting, hardening of the stalks and spikes and development of the prickle like structures on them.

2.6 Nematode problems in the nursery of horticultural crops

Nematode infested nursery is the important threat in horticultural crops, and most of these crops are raised in the nursery. Infested nursery is the source of dissemination of nematode in the horticultural field. Once the nematodes have introduced in the main field it is impossible to get rid from these worms. Seedling raised in nursery are infested by several nematodes produced in Haryana and other part of the country. Nematodes attack on seedlings form galls and stunted growth are the indicators of nematode damage in nursery (**Figure 3**). *Meloidogyne* spp. (root-knot nematodes) and *Rotylenchulus reniformis* (reniform nematode) are the most damaging nematode among the plant parasitic nematodes in nursery.

2.7 Nematode-pathogen interaction: a real threat

Nematodes associated with other secondary micro-organism increased the losses' manifolds. Phyto parasitic nematodes provide avenues for secondary pathogens viz., fungi, bacteria, viruses, etc. and favor the establishment on host. Nematode alter the host physiology for colonization of secondary microbes. However, nematode

Figure 3.
A. Nematode infestation in the nursery B. guava nursery infested with root knot nematode (source: Original photos).

Figure 4.
Root infested with Meloidogyne *spp. and wilt fungus (source: Original photos).*

themselves are capable of causing significant losses in the crops, their connotation with secondary microbes preponed the disease and compound the damage. The rotting fungi (*Rhizoctonia bataticola*, *R. solani*) and wilt fungus, *Fusarium oxysporum* has been reported in compounding the disease severity by nematode [12]. Combined infection of nematode and fungal pathogens leads to sever rotting and death of plants (**Figure 4**). Sometimes, presence of both pathogens, nematode and secondary pathogens breaks the resistance in resistant cultivars of plants [13].

3. Nematode management in horticultural crops

Aim of nematode management is to reduced crop losses in yield and quality of crops and manage the nematode population below economic threshold level.

4. Preventive measures

4.1 Soil testing for nematode mandatory before establishing polyhouses/ net-houses

Soil testing is mandatory for the farmers before erection of the polyhouses, green house, net houses and orchards.

4.2 Nematode-free planting materials

Infested planting material is the foremost important means of spreading of nematodes in horticultural crops. At present, there is no chemical or non-chemical methods are available that resolve the nematode problem in protected cultivation once introduced into the crops. Nematode-free planting material that produced on soilless substrates are increasingly used to exclude soil borne species of nematodes, but also to promote the plant establishment and crop production.

4.3 Hot-water treatment

Hot water treatments was successfully used to manage plant parasitic nematodes spread through planting material in some crops like strawberry, citrus. In this technique temperature and time combination is very important otherwise the planting material may become unfit for transplanting.

5. Curative measures

5.1 Sanitation

Rapid destruction of infested planting material and weeds from the field that can help in minimizing the nematode population and prevent further spread of nematodes. Entry points and equipment used in protected structures should be sanitized. Using clean water also helps in minimizing the further spread of pest.

5.2 Tillage and soil solarization

Soil solarization in the hottest part of the year by using transparent polyethylene plastic (LLDP 25 μm) 6–8-week period after harvesting of the crop are very helpful in reducing nematode population in the soil. A plowed field with moist soil covered with polyethylene sheet help in raising the soil temperature and is lethal for the soil borne pests.

Soil tillage in the months of May–June in northern Indian conditions for two to three times also helps in reducing the nematode populations. Soil solarization or tillage is dependent on hot weather and fallow soil with the crop-free period of 4–6 weeks is necessary which may not be economically feasible to the growers.

5.3 Crop rotation

Continuous monoculture of the susceptible crop has tremendously increased the pest problem including plant parasitic nematodes. In such circumstances, crop rotation with the non-host crop is one of the good options to reduce nematode build up. Popularity of this method depends upon the availability of suitable non-host crop. The length of rotation is however related to the magnitude of the initial nematode population and the rate of population decline during rotation. In protected structures, rotating susceptible crops with poor or non-hosts, trap crops, antagonistic crops or biofumigants such as brassicas, that fetch profit to grower, can be an alternative.

5.4 Resistant variety or grafting on resistant rootstock

Resistant variety or grafted resistant root-stock is one of the convenient methods of nematode management in horticultural crops. In India, still there is no released resistant variety to nematodes in protected conditions. However, in India work is restricted to evaluation of germplasm against plant parasitic nematodes. Grafting of commercially important crop genotypes on nematode resistant rootstocks (eg. scion of commercial tomatoes on wild eggplants root stocks, *Solanum toxicarium*, *S. sisymbriifolium* and *S. torvum*) is an efficient choice for management of *Meloidogyne* spp. in vegetables under protected cultivation system [14]. Susceptible but agronomically desired arabica scion on resistant Robusta root stock of coffee have been effective against *Pratylenchus coffee*.

5.5 Organic amendments

Use of organic amendments is a good option to reduce nematode build up as well increase plant tolerance by raising nutrient status. Both edible and non-edible oil cakes are used for suppressing nematode population in soil. Organic amendments manage the nematode population by release of toxic compounds during decomposition, improve soil fertility increased the plant vigor, tolerance and promoting antagonistic microbial activity.

5.6 Biological control

Recently, bio-agents has paid much attention due to popularity in suppression of nematode population as well as environmentally sound option for management. Bio-agents such as egg parasitic fungi – *Purpureocillium lilacinum*, *Pochonia chlamydosporia*, antagonistic fungus – *Trichoderma viride*, *Trichoderma harzianum*, bacterial parasites – *Pseudomonas fluorescence*, *Pasteuria penetrans* and Mycorrhizal fungus, *Glomus fasciculatum* has to be used in nematode management. Amendments (neem cake, vermicompost or FYM) enriched with bio-agents was found potential for management of plant parasitic nematodes in polyhouses [15].

5.7 Chemical nematicides

New nematicides such as 3-F nematicides that all having Trifluoro group in their molecular structure (fluensulfone, fluopyram, and fluazindolizine) are very effective against plant parasitic nematodes but till now they are not registered for use in many crops in India.

At present, there are no conventional fumigant or non-fumigant nematicides registered for greenhouse use in India. Thus, protected cultivation growers are dependent on other IPM practices such as exclusion, sanitation, nematode resistant plant varieties when available, and other cultural and biological means of nematode management.

6. Integrated nematode management

Pre-plant application of fumigants followed by neem cake fortified with bio-agents suppressed the plant parasitic nematode populations under polyhouses. Nonetheless, integrated nematode management in protected structures have certain limitations such as non-host availability, crop preference, plant parasitic nematode incidence at different geographical locations. Keeping in view of this, suitability of integrated management practices with general farming practices has to be remolded according to local conditions.

7. Conclusions

Farmers/growers neglect the nematode diseases in the crops because of hidden nature and non-diagnostic symptoms produced by nematodes on crops. Plant parasitic nematodes associated with secondary microbes present in the rhizosphere form disease-complexes and increase crop damage. Changes in the cropping system, also changes the nematode community structure and there by nematode problem. A menace of plant parasitic nematodes is increasing day by day under protected conditions, orchards, plantation crops, mushroom houses and field crops.

Clearly, recent research on nematode management in polyhouses has focused primarily on plant parasitic nematode and soil borne fungal pathogens. Nematodes are extremely diverse habitats, ubiquitous and vary enormously in their lifestyles, we shall gain knowledge of nematode behavior and biology of nematode taxa for the comparative approach under more realistic environmental conditions. Thorough knowledge of tolerance, behavioral and sensitivities of nematodes to various environmental extremes is an important for fundamental science and management.

Acknowledgements

The authors greatly acknowledge the help of Dr. R S Kanwar, Prof. and Head Nematology, CCSHAU, Hisar for critically reviewing the manuscript and giving suggestions for its improvement.

Conflict of interest

The authors declare no conflict of interest.

Author details

Saroj Yadav* and Jaydeep A. Patil
Department of Nematology, College of Agriculture, CCS HAU,
Hisar, Haryana, India

*Address all correspondence to: sarohau29@gmail.com

References

[1] Altieri Miguel A, Fernando R Funes-Monzote, Paulo Petersen: "Agroecologically efficient agricultural systems for smallholder farmers: Contributions to food sovereignty." Agronomy for sustainable development 32.1. 2012; 1-13.

[2] Anonymous: National Horticulture Board, Government of India. 2018.

[3] Singh R P: "Horticultural (high value agricultural) crops diversification in eastern India: II—Employment opportunities and income generation strategies." International Journal of Innovative Horticulture 2.1. 2013; 28-43.

[4] Piali H, Pati S: "A need for paradigm shift to improve supply chain management of fruits & vegetables in India." Asian Journal of Agriculture and Rural Development 1.393-2016-23908. 2011; 1-20.

[5] Kumar V, Khan M R, Walia R K: Crop loss estimations due to plant-parasitic nematodes in major crops in India. 0National Academy Science Letters. 2020; 43: 409-412.

[6] Patil J, Goel S R, Yadav S: Effect of *Meloidogyne incognita* and *Fusarium oxysporum* f. sp. *cucumerinum* on cucumber grown under protected cultivation. Journal of Entomology and Zoology Studies. 2017; 6: 1004-1007.

[7] Poornima K, Suresh P, Kalaiarasan P, Subramanian S, Ramaraju K: Root knot nematode, *Meloidogyne enterolobii* in guava (*Psidium guajava* L.) a new record from India. Madras Agricultural Journal. 2016; 103: 359-365.

[8] Evans K, Stone A R: A review of the distribution and biology of the potato cyst-nematodes, *Globodera rostochiensis* and *G. pallida*. Pans. 1977; 23: 178-189.

[9] Seshadri A R: Chemical control of potato nematodes. nBuoG PmcD ATA SHEET PN-AAJ-076 AHIo-0210-oo0, 1978; 173.

[10] Chandel Y S, Bhadu S S, Salalia R, Thakur S, Kumar S, Somvanshi V S, Mukherjee A Walia R K: Prevalence and spread of potato cyst nematodes, *Globodera* spp. in northern hilly areas of India. Current Science. 2020; 118: 1946.

[11] Buckowski T: Eelworm control. A practical method for the sterilization of casing soils. Mushroom Science. 1964; 6: 485-492.

[12] Patil J, Kumar A, Goel S R: Incidence of plant-parasitic nematodes associated with Polyhouses under protected cultivated in Haryana. Environment and Ecology. 2018; 35: 1870-1873.

[13] Sidhu G, Webster JM: Predisposition of tomato to the wilt fungus *Fusarium oxysporum* by the root-knot nematode *Meloidogyne incognita*. Nematologica. 1977; 23: 432-442.

[14] Louws F J, Rivard C L, Kubota C: Grafting fruiting vegetables to manage soilborne pathogens, foliar pathogens, arthropods and weeds. Scientia Horticulturae. 2010; 127: 127-146.

[15] Sabir N, Walia RK: Management of nematodes in protected cultivation with short notes on key pests. https://www.researchgate.net/publication/322676529. October, 2017

Section 2

Management of Plant-Parasitic Nematodes

Chapter 3

Role of Microbial Enriched Vermicompost in Plant-Parasitic Nematode Management

Sunil Kumar, Ranjit Kumar and Pankaj Sood

Abstract

Earthworm causes increase in availability of soil organic matter through degradation of dead matters by microbes, leaf litter and porocity of soil. Vermicompost is a non-thermophilic biodegradation process of waste organic material through the action of microorganism with earthworm. Vermicompost is rich in many nutrients including calcium, nitrates, phosphorus and soluble potassium, which are essentially required for plant growth. Different plant growth hormones like gibberellins, auxins and cytokinins are present in vermicompost, which has microbial origin. Nematodes are mostly small, colorless and microscopic organisms which remain under soil, fresh or marine water, plants or animals, and act as parasite in different conditions, while very few have direct effect on human. The nematodes which are parasitic on plants use plant tissues as their food. They have well developed spearing device, like a hypodermic needle called stylet. It is used to penetrate host cell membrane. Management of plant-parasitic-nematodes therefore is necessary and several means are adopted. Of which, use of bio-chemicals and organic compost have shown encouraging results and proved to be potential in suppressing the nematode population. Vermicompost plays an important role of soil fortification on growth characteristics, such as length, weight, root, shoot branches, number of leaves and metabolism of host plant against nematode infection. Vermicompost fortified plants showed increment in sugar, protein and lipid over untreated control. Increment of these metabolites helps treated plants to metabolically cope up the infection and promotes excessive plant growth. The vermicompost caused the mortality of nematodes by the release of nematicidal substances such as hydrogen sulfate, ammonia, and nitrite apart from promotion of the growth of nematode predatory fungi that attack their cysts. It favours rhizobacteria which produce toxic enzymes and toxins; or indirectly favors population of nematophagous microorganisms, bacteria, and fungi, which serve as food for predatory or omnivorous nematodes, or arthropods such as mites, which are selectively opposed to plant-parasitic nematodes.

Keywords: Vermicompost, nematode, nematophagous, *Meloidogynae*

1. Introduction

The term vermicompost is derived from a latin word "vermes" meaning "worms" and the process of composting of organic material using earthworms is known as vermicomposting. Earthworms directly influences the microbial community of soil and it maintains normal chemical and physical properties ofsoil, due to which it is popularly called the "farmer's friend".

Earthworm causes increase in availability of soil organic matter through degradation of dead matters by microbes, leaf litter and porocity of soil. Vermicompost is a non-thermophilic biodegradation process of waste organic material through the action of microorganism with earthworm. The product produced through vermicompost is highly fertile, very finesoilparticles with marked porosity, adequate aeration, low C: N ratios and high water-holding capacity [1]. The termed "drilosphere" is coined for microflora and microfaunain soil influenced by earthworms [2].

Due to decrease in land availability for cultivation, waste disposal and exponential increase in human population there is urgent need to improve crop production and waste disposal mechanism. Crop intensification has led to huge use of chemical fertilizers and pesticides which plays key role in ecological disturbances by destroying natural predators of crop pests, plant growth-promoting bacteria and other soil micro/macro flora and fauna. These pesticides pollute environment very adversely, necessitating demand for safe organic farming to protect us from adverse effect of these pollutants. Organic waste composting is a technique which converts organic wastes into useful composts, which could be used as biofertilizer for sustainable agriculture growth. Conventional composting through microbes is a thermophilic process, in which many microbes are lost due to excess temperature emitted during the composting process. While vermicomposting is a mesophilic process, which conserves all microbes and earthworm associated with it to provide associated beneficial effect for degradation of organic matter by preserving the diverse community of all beneficial microflora. Vermicompost provides more biologically active and nutritive biofertilizers in soil as earthworms transform different organic waste material into useful vermicompost material by grinding, churning and digesting these substancesin association with microbes which is essential in biogeochemical processes [3]. Earthworms enhance the beneficial microbes and suppresses harmfulmicrobes to convert different infectious hospital wastes into risk-free materials [4].

Vermicomposts are rich in many nutrients including calcium, nitrates, phosphorus and soluble potassium, which are essentially required for plant growth [5]. Different plant growth hormoneslike gibberellins, auxinsand cytokinins are also present in vermicompost, which has microbial origin.

Nematodes are mostly small, colorless and microscopic organism remain under soil, fresh or marine water, plants or animals They act as parasite in different conditions, while very few has direct effect on human. Almost 50 percent nematodes are living in marine environment while about 25 percent of the nematode species live in soil and fresh water feeding on different decomposer organism including bacteria and fungi, many small invertebrates and organic waste Only 15 percent of the nematode species are parasitic in nature, infecting animals, ranging from small insects, many invertebrates and man. Only almost 10 percent nematode species are plant parasite in nature.

The nematodes which are parasitic on plant use plant tissues as their food through well developed spearing device like a hypodermic needle called style used to penetrate host cell membrane. Plant-parasitic nematodes release an enzyme into a host plant cell through stylet for partial digestion of cell content before entry into gut. Nematode causes injury to plants in two ways involving their feeding mechanism. Few nematodes are ectoparasitic which utilizes different plant tissues outside the plant for their food, while few nematodes are endoparasitic which utilize inner part of plant tissues as their food. Few nematodes are migratory known as foliar nematodes which utilizes the leaves and buds of ferns, chrysanthemums, strawberries and many other ornamentals as their food. Foliar nematodes cause death of buds, distortion of leaf and formation of dark-brown to yellow lesions between major veins of the leaves.

Management of plant-parasitic nematodes therefore is necessary and several means are adopted. Of which use of bio-chemicals have shown encouraging results and proved to be potential in suppressing the nematode population. Vermicompost plays an important role of soil fortification on growth characteristics, such as length, weight, root, shoot branches, number of leaves and metabolism of host plant against nematode infection. Vermicompost fortified plants showed increment in sugar, protein and lipid over untreated control. Increment of these metabolites treated plants and metabolically cope up the infection and promote excessive plant growth. Use of Vermicompost as fertilizer also helps in suppression of plant diseases and pest as it provides better nutrient availability and greater strength, immunity, and resistance against infection. Compost and vermicompost are effective in eliminating root-knot nematodes (*Meloidogyne incognita*) in tomato plants. Almost 40 species of bacteria and 22 species of fungi were identified in soil treated with vermicompost, of 40 bacterial species majority were found under the genus like *Azotobacter*, *Bacillus*, *Rhizobium*, *Pseudomonas*, *Beigerinicka* and *Enterobactor* and in fungal the genus such as *Aspergillus*, *Rhizopus*, *Pencillum*, *MucorCladosporium*, and *Fusarium* were commonly found [6]. Arancon et al. [7] had also observed a reduction in plant-parasitic nematodes following the application of vermicompost.

2. Vermicompost

2.1 Nutritional composition

The nutrient content obtained from vermicompost directly depends on the constituent of waste material where it feeds. It enhances levels of different material in casted soil than available mineral concentration due to microbial activity in its gut [8]. According to reports of Hand *et al.* [9] the earthworms enhance nitrogen mineralization in the soil, consequently resulting in more availability of nitrate in the soil. The vermicompost is also involved in reduction of organic carbon and carbon nitrogen ratio than in the normal composts. The combined earthworm and microorganism action lowered causes loss of different organic matter from the soil substrates as CO_2 introduces 20–43% of total organic carbon material in soil the completion of vermicomposting period. Vermicompost also contains all essential nutrients including nitrates, phosphate, exchangeable calcium and soluble potassium which are quickly absorbed by plants (Edwards, 1998; [10]). Also observed more micro and macro nutrients in the vermicompost which are rich in the earthworm casts.

2.1.1 C/N ratio

The carbon and nitrogen (C/N) ratio is most important parameter during composting process which clearly indicates about the decomposition rate. Plants are able to take mineral nitrogen in the form of nitrates, only when carbon and nitrogen ratio falls below 20 [11]. The proper ratio of carbon and nitrogen is therefore required for the proper plant growth. Earthworms cause reduction in carbon level thereby increasing the nitrogen content in fresh organic matter.

2.1.2 Nitrogen

Nitrogen is very essential constituent of all amino acids and protein. Deficiency of nitrogen directly decreases the growth of plants leading to chlorosis, stunted and slow growth. According to Hand *et al.* [9] mineralization with nitrogen was highly facilitated in earthworm presence and it leads to deposition of nitrate in the soil.

2.1.3 Phosphorus

Earthworms activity causes increase in total phosphorus concentration in soil in comparison to the food source available in soil. This clearly indicates that the vermicomposting causes increase in phosphorus level through the mineralization of phosphoric organic compounds [12, 13].

2.1.4 Iron (Fe)

Iron (Fe) is also an important element required for growth and productivity of all plants. Only very trace amount of iron is required in comparison to other minerals by plant like carbon, oxygen, hydrogen, nitrogen, phosphorus, sulphur and potassium for proper plant growth. The iron functions like a cofactor, as it has a catalytic site for many essential enzymes activity which are even required for chlorophyll synthesis.

2.1.5 Magnesium (Mg)

It is a important component used in formation of chlorophyll, which play vital role in photosynthesis. It is also required for carbohydrate metabolism and acts as enzyme activator in nucleic acid synthesis. Magnesium serves as a carrier of phosphate compound in plants and also supports uptake of many essential elements into plant. It enhances production of oils and fats through the translocation of carbohydrates.

2.1.6 Manganese (Mn)

Manganese (Mn) plays vital role in nitrogen assimilation by, as enzyme activator. It is very important constituent of chlorophyll. Low plant manganese usually causes leaves to turn yellow due to reduced chlorophyll content. Organic soils usually contain intermediate amounts of manganese.

2.1.7 Zinc (Zn)

Low presence of zinc leads to high yield of crops. Zinc efficiency has been reported in many enzymatic activities of plants [14]. Zinc utilization mechanism in plant tissue is most important mechanism of zinc in plant tissues. Heavy metal bioaccumulation study showed that increased duration of vermicompost concentration of Zn and Cu decreases soil [15].

2.2 Role of vermicompost in plant growth promotion

Wide variety of plant species grows effectively in vermicompost rich soil, including many horticultural crops like tomato, cauliflower etc. [16], aubergine [17], garlic, pepper [18], strawberry, green gram and sweet corn [19]. Vermicompost is also very much effective on enhance production of many medicinal plants rich in aromatic compounds [20], cereals such as rice and sorghum [21], fruit crops such as papaya and banana, and ornamentals like geranium [22], petunia, marigolds and poinsettia. Effect of vermicompost was also observed in forest trees including eucalyptus, acacia and pine tree [23]. Vermicompost are very beneficial and used as a partial or total substitute for chemical fertilizer in agriculture and artificial greenhouse potting media. Likewise, few studies show that water-extracts obtained from vermicompost, vermiwash were used as foliar sprays, which enhances growth of tomato plants [24], strawberries and sorghum. Vermicompost also stimulates seed germination in green gram and other plant species [25], tomato plants [26], pine trees and petunia. Vermicompost are used effectively for vegetative growth of leaf, stimulating growth of root and shoot [27]. These effects cause increase in root branching and leaf area and alterations in morphology of seedling plant [28]. Vermicompost stimulates flowering in plants, increasing flowers produced [29], and increase in fruit yield [30].

2.3 Bacterial diversity associated with earthworms

A variety of bacterial species have been reported associated with earthworms/vermicompost though the bacterial species varied with its isolation site including soil, intestine, and excrements. Almost 43 bacterial species were isolated from earthworm intestines and 25 obtained from fresh excrements of which, 9 were common. Among 40 bacteria species isolated from soil and intestine, 13 were shared species; 9 were gram-positive, and 6 Bacillus species were spore-forming. Comparison of soil and excrements bacteria revealed similarity of only 6 isolated species, of which three species were gram-positive and three species were gram-negative. *Brevundimonas diminuta* (α-Proteobacteria), *Kocuria palustris* (Actinobacteria) and D. acidovorans (β-Proteobacteria), were isolated from all three substrates. Comparison of bacteria isolated from the intestine of *Aporrectodea caliginosa*, *Lumbricus terrestris*, and *Eisenia fetida* earthwormsrevealed that the highest number of 43 bacterial taxa was isolated from *A. caliginosa* digestive tract; while from *L. terrestris* and *E. fetida*, 22 and 21 taxa were isolated respectively. Few members of bacteria were isolated from all earthworm species, which includes Bacteroidetes (classes Flavobacteria and Sphingobacteria), Actinobacteria, Proteobacteria (classes α-, β-, γ-) and Firmicutes (class Bacilli). Five bacterial species isolated from earthworm exhibited relatively low similarity between the sequenced 16S rRNA gene fragments (approx. 1490 nucleotides) and the genes of known bacterial taxa (93–97%), which includes Ochrobactrum sp. 341-2 (αProteobacteria), *Sphingobacterium* sp. 611-2 (Bacteroidetes), *Massilia* sp. 557-1 (β-Proteobacteria), *Leifsonia* sp. 555-1, and a Microbacteriaceae, isolate 521-1 (Actinobacteria).

Micromycetes were observed in digestive tracts of fasted earthworm species. The incubation temperature had no effect on the number of fungal CFU isolated from the intestines. Fungi isolated from the earthworms after 20 days of starvation, are *Bjerkandera adusta* and *Syspastospora parasitica* identified by light-colored sterile mycelia, as well as *Geotrichum candidum*, *Alternaria alternata*, *Acremonium murorum* (*A. murorum* var. *felina*), *A. versicolor*, *Aspergillus candidus*, *Rhizomucor racemosus*, *Mucor hiemalis*, *Cladosporium cladosporioides*, Fusarium (F. *oxysporum*, *Fusarium* sp.), and *Penicillium* spp.. The density of fungal colony in the air dry intestine was 103–104 CFU; this value is very close to the fungal populations density in soil mineral horizons. These fungi are most resistant to the conditions within earthworm digestive tract.

2.4 Role of vermicompost in nematode control

The application of vermicompost resulting in reduction of free-living nematodes populations owing to, its adverse effects on these nematodes. The management of plant-parasitic nematodes is very difficult in comparison to management of other insect pests and pathogens. The plant-parasitic nematode generally resides in soil and attacks the underground parts of plants. While cyst nematode management faces a unique challenge owing to hard protecting cyst wall protects egg of gravid females. Prevention is the most common economical control method, because once any plant is parasitized by nematode, it is essential to destroy host for killing worm effectively. At present chemical nematicide is commonly used in controlling different types of plant-parasitic nematodes in the soil [31]. Frequent treatment of soil with different chemical is dangerous and adversely affects soil organisms, environment, as well as animal and human health. Gabour *et al.* [32] observed inhibitory effect of vermicompost application on the populations of the plant-parasitic nematode *Rotylenchulus reniformis*. In addition to vermicompost, recent studies have shown that the application of vermicompost tea has the potential to control plant-parasitic nematodes. In this sense, Edwards et al. [33] studied a significant suppression in the number of galls caused by *Meloidogyne hapla* in tomato when the plants were subjected to aerated vermicompost tea. The effects of vermicompost are likely on nematodes due to the mortality of nematodes by the release of nematicidal substances such as hydrogen sulfate, ammonia, and nitrite produced [34]. promotion of the growth of nematode predatory fungi that attack their cysts [35]; favoring of rhizobacteria that produce toxic enzymes and toxins [36]; or indirectly by favoring populations of nematophagous microorganisms, bacteria, and fungi, which serve as food for predatory or omnivorous nematodes, or arthropods such as mites, which are selectively opposed to plant-parasitic nematodes [37].

3. Mechanisms that mediate nematode control

3.1 Crop rotation

It reduces many soil-borne diseases and improves soil for agriculture. Many nematodes can reproduce, grow and survive on selected plants and not able to grow on other crops and hence die with practice of crop rotation. Repeatedly growing of single crop in particular field will enable any organism to reproduce successfully and increase their number. While introduction of crops which does not support nematode growth will prevent reproduction and growth of nematode and allow natural mortality factors to act on these to reduce their numbers. Through the planned rotation of crop and sequential alteration of crop, it is possible to reduce excessive growth of all pests of all of the major agriculture crops. Hairy indigo would reduce numbers of sting and rootknot nematodes and can be planted as a summer crop in between other crops. Pangola digitgrass a common agriculture crop of West Indies and Florida, which control burrowing and root-knot nematodes in vegetable lands. Use of crop rotation usually provides multiple benefits including mineralization of soil as well as effective pest control in agriculture field.

3.2 Crop root destruction

Through destruction of root whole pest colony which resides on root are destroyed, and leads to decline in number of nematodes through increased mortality. This can stop nematode reproduction and should encourage their decline through normal mortality. Crop root systems are reservoir for many soil-borne diseases and nematodes. These small insects and nematodes can multiply on root system of crop whenever it will remain alive. Almost 10 folds increase in nematode concentration were observed when soil temperature were high. Even when soil temperatures are declining, at least one additional generation of nematode were found. It was very good practice in nematode management to destroy root of previous crop to prevent growth and reproduction of nematode.

3.3 Flooding

Flooding the agriculture land was also used to reduce numbers of soil infesting pests including plant-parasitic nematode. It is done through regular maintenance of high water level in field for many weeks in controlled manner. This high water level is maintained in field for two or three weeks followed by drying the soil and flooding again for two to three weeks is more effective way of controlling plant-parasitic root nematode. Flooding generally kills root nematodes by inhibition of nematode parasite with interaction to host plants for longer period.

3.4 Fallowing

It is a process in which a field is left without any type of vegetation and plants for longer period; it leads to starvation of nematodes or other pests in absence of vegetation. Most soil pests and nematodes were decreased due to lack of food in the form of host plants. The field must be regularly cultivated to prevent growth of different weeds and it leads to proper cycling of drying and heating to different layers of soil.

3.5 Plant resistance

Many plants are resistant to different types of pests, And their use in agriculture field is most effective and less expensive way of pest control strategy. But this method requires detailed knowledge on various resistant plants and pest categories and situation which does not support pest survival, but most of nematode resistant crop has resistant for only few nematode species and it would not be completely resistant to all species of nematode.

3.6 Biological control

Many biological agents like bacteria and fungi are nature well known enemies of nematodes. These do not support growth of nematode species when concentration of these bacteria and fungi are high. Many scientific studies on nematode are able to reduce nematode population with the help of these bacteria and fungi under laboratory conditions. But at field levels this is emerging field of research and success rate are not very high. However, the use of organic materials to the soil has been found reported to increase the availability of food for fungivorous and bacteriophage nematodes, increasing the competition between them with other groups.

4. Nematode associated with agricultural crops

Nematode species varies greatly in different countries. Few nematode species are cosmopolitan, likespecies of *Meloidogyne* while many are geographically restricted to particular region e.g. different species of *Heterodera, Globodera,* etc. *Nacobbus*species are highly specific attacking only carrots. Some crops are infected with very few species of nematode pests while others are infected with wide range of nematode species. Crop like rice, maize and sugar cane are infected with variety of plant-parasitic nematodes. Common plant-parasitic nematodes associated with different agricultural crops are: *Meloidogyne, Heterodera, Globodera, Pratylenchus, Radopholus, Rotylenchulus, Tylenchorhynchus, Xiphinema, Longidorus, Paralongidorus, Aphelechoides, Ditylenchus* etc.

5. Nematode management through microbial biogents

Nematode mainly attack underground parts of plant, due to which management of nematode are very difficult in comparison to other plant parasites [38]. At present synthetic nematicides are frequently used for management of plant-parasitic nematodes [39]. Although, nematicides are very efficient and fast acting, but are relatively unaffordable to many small scale farmers.

Application of organic amendments is one of the best practised alternative strategies for the management of plant-parasitic nematodes in the soil [40]. Organic amendments have shown beneficial effect on soil physical conditions, soil nutrients, and soil biological activity, thus improving plant health and reducing colonies of plant-parasitic nematode [41]. Integrated pest management (IPM) uses different strategies for the management of plant-parasitic nematode but the biological control would be the most effective and economical way of nematode management. Different groups of beneficial bacteria have been utilized for the management of plant-parasitic nematode in soil. Various fungi such as *Aspergillus*, *Paecilomyces*, *Trichoderma*, *Verticillium*, *Pochonia*, *Fusarium* and *Penicillium* have been reported to cause juvenile mortality and egg inhibition of nematodes. An increase in nematicidal potential of microorganisms were observed when such bacteria, fungi or other biocontrol agents are integrated with either organic amendments or nematicides for integrated control of nematodes [42, 43].

6. Ecological and economical importance of biomanagement

The nematodes can survive in different environments including aquatic (such as fresh water, estuarine and marine water), terrestrial (as free living in the soil) and parasitic (either endoparasites and ectoparasites of animals and plants). Pokharel, and Larsen [44] and Pokharel., *et al.* [45], reported that soil nematodes are very important in protecting the organic nature of soil, Phytoparasitic nematodes on feed tissue of plants and reduce the growth and productivity of infected plants. Soil nematodes assist colonization of microbial substrates and nutrients mineralization in the soil. Metabolism in nematodes produce important nutrients like nitrogen and vitamins which speed up bacterial growth in the soil. Many nematodes feed on bacteria, fungi and protozoa within soil and acts like predatory or omnivorous nematodes it would improve cycling of nutrients and causes slow release of nutrients into soil. The free-living nematode in soil enhances mineralization of nutrient in soil. These nematode groups also feed different plant pathogen and few soil microbes including plant pathogens such as bacteria and fungi. Free-living nematodes can protect system crop by protecting nature of soil. The nematodes which attack insect pests are useful biological insecticide [46].

According to data of the American Phytopathological Society, nematodes have great economic benefits of both harmful and useful effect, most plant nematodes has a sharp needle-like structure found in mouth part called stylet. They is cause more than 15percent loss of crops per annum world-wide, equal to almost US$78 billion. Majority of plant feeder nematodes found in the soil, feed on plants and reduce water and nutrient absorbed by the plants root, reducing their drought resistance ability. Some other nematodes transmit disease causing organisms like viruses to plants while feeding. Large number of nematode species cause decomposition and recycling of nutrient by release of relevant nutrients for the plant growth.

From more than 4000 described plant-parasitic species of nematodes, only some cause economic losses in crops. Some of the major genera of phytoparasitic species of nematodes causing crop losses are *Xiphinema, Rotylenchulus, Pratylenchus, Meloidogyne, Hoplolaimus* and *Heterodera* [47]. The majority of soil nematodes are present in the rhizosphere of plant root area in the soil surrounding the root of plant where microbiological activity is exceptionally high.

7. Enrichment of beneficial microorganisms in vermicompost

7.1 Enrichment of vermicompost with bacteria

Earthworm's gut microflora has high ability to increase plant nutrient availability. Earthworms highly influence the soil dynamics and chemical processes, by adding its litter and affecting the soil micro-flora activity [8]. Earthworms and microorganisms interaction seem to be very complex. Earthworms excretes plant growth-promoting substances and making soil fertile. Pseudomonas oxalaticus an oxalate-degrading bacterium was isolated from intestine of different species of earthworm and *Streptomyces lipmanii*from actinomycetes group was identified in the gut of *Eisenia lucens*. Scanning electron micrographs showed presence of endogenous microflora in guts of earthworms, *L. terrestris* and *Octolasion cyaneum*. Gut of *E. foetida* contained various anaerobic N2-fixing bacteria such as C. *Beijerinckii, Clostridium butyricum* and *C. paraputrificum*.

7.2 Enrichment of vermicompost with fungi

A total of 194 fungal entities comprising 117 mitosporic fungi, 45 ascomycetes, 15 zygomycetes, 14 SM morphotypes and three basidiomycete morphotypes were reported from the vermicompost. Mitosporic fungi including the ascomycetes in their anamorphic state are the most dominat. The thermotolerant fungus, Scedosporium state of *Pseudallescheria boydii* also display a significantly high load in vermicompost, However *Penecillium* and *Aspergillus* showed highest load in vermicompost.

8. Enrichment of vermicompost and agriculture benefits

Vermicomposting is biotransforming process, stabilizing waste organic materials into humus by joint activity of microorganism and earthworms. Earthworms excrete casts which are partially digested waste materials, commonly known as castings or vermicast, and are homogeneous in composition of minerals than the source waste material. Vermicompost has very least levels of contaminants, and contains increased amount of minerals, plant hormones, symbiotic microbes and organic acids including fulvic and humic acids. Vermicomposting is a process of compost production by breeding, growing and maintaining earthworms population in soil. The earthworms cause biooxidation of waste by relentless turning, aeration, and fragmentation resulting in formation of homogeneous and stabilized humus in soil, which is useful for plant thus used as manure in agriculture field. Vermicomposting is very effective for maintenance of biodegradable household waste and Municipal Solid Waste at many places. Aerobically incubated extract of compost are now in high demand commercially for agriculture work. As being are rich in carbohydrate and a protein source. It is known as 'compost teas' which is microbially very rich. The casting by earthworms consists of many nutrient including Nitrogen, Phosphorus, Potassium, Calcium and Magnisium.

9. Conservation of microbial biogents

9.1 Bacteria

Bacteria are generally found in very diverse habitat including marine water, fresh water, soil and compost piles. Many bacteria are found in gastrointestinal tract of animal system. Few bacteria also reside in oxygen deficient conditions like in flooded soil. While most bacteria required well aerated soil. Many of bacteria grow and reproduce very rapidly in acidic and neutral soil conditions. Bacteria are first decomposer found in soil which initiates process of decomposition of different material in it. Through the process of decomposition bacteria makes different minerals available to plants. Phosphorus is also dissolved by bacteria and plants can utilize this dissolved phosphorus easily for their growth. Nitrogen fixing bacteria fixes nitrogen in soil for plants. Plants require large amount of nitrogen in agriculture soil for proper growth. It is well known fact that nitrogen present in atmosphere is neither consumed by animals nor plants for their nutrition and growth. Few nitrogen fixing bacteria has ability to convert these nitrogen gas into nitrate which is easily absorbed by plants. Plants use this nitrogen compound to form different types ofamino acids and proteins. This process of formation of nitrate compound through free nitrogen is called *nitrogen fixation*. Nitrogen-fixing bacteria generally reside in root nodules of plant to form mutually beneficial symbiotic associations with plants. Rhizobium bacteria reside in root nodule of different leguminous plant and fixes nitrogen present in air effectively while these bacteria uses sugars of plants for their energy source. Bacteria in alfalfa field can fix many hundreds pounds nitrogen per acre per year.

Pea plant fixes very less amount of nitrogen in field, it accounts for only 30 to 50 pounds per acre. Large molecules of lignin were broken down into very smaller in size through actinomycetes. Lignin is a complex and large molecule found in plant tissue, it protect cellulose from decomposition, bacteria acts on it and degrade it in to simpler form during the process of decomposition. Earthworms can also facilitate the dispersion of microorganisms by the excretion of their spores in the coprolites. However, the dispersion of nematophagous fungi by earthworms might be responsible for the reduction of the nematode populations in the substrates. The mechanisms by which vermicomposts and their aqueous extracts suppress plant-parasitic nematodes after application to soil, are speculative. Larger predator–prey populations can also contribute to lower densities of plant-parasitic nematodes in vermicompost-treated soils [48]. Vermicomposts can increase the numbers of predatory or omnivorous nematodes or arthropods such as mites that selectively prey on plant-parasitic nematodes [48, 49]. Vermicomposts can promote the growth of nematode-trapping fungi and fungi that attack nematode cysts and may thereby influence the populations of plant-parasitic nematodes [35].

9.2 Fungi

Fungi are also important constituent of plant microorganism. Many fungi produce a number of antibiotics. Fungi also initiate the decomposition of waste as well as fresh organic residues. They act on surface of material, making it soft and available for other microorganism for initiating the decomposition of organic material. Decomposition of lignin also require fungal activity followed by bacterial decomposition process.

9.3 Algae

Many algae, like crop plants, make sugars with the use of sunlight and carbon dioxide, which is used for energy need and formation of other complex molecules. Flooded soil and rice paddy field are rich in many species of algae. These algae grow on surface of wet soil and form mutually beneficial relationship with other organism for enhancing nitrification and mineralization process in soil. It also shows formation of lichens in agriculture field.

9.4 Protozoa

Different species of protozoa use a variety of means for increased productivity of soil. Protozoan's feed on bacteria, fungi and other protozoa and waste materials. Protozoa acts like secondary consumers of organic materials. Protozoa consuming nitrogen rich organisms and excreting wastes rich in nitrogen element this is believed to be responsible for mineralizing much of the nitrogen in agricultural soils.

Author details

Sunil Kumar[1*], Ranjit Kumar[1] and Pankaj Sood[2]

1 Department of Animal Science, School of Life Science, Central University of Himachal Pradesh, Dharamshala, H.P., India

2 CSK Himachal Pradesh Krishi Vishvavidyalaya, Palampur, India

*Address all correspondence to: sunilibes@gmail.com

References

[1] Domínguez, J. & Edwards, C.A. (2010). Relationships between Composting and Vermicomposting: Relative Values of the Products, In: Vermiculture Technology: Earthworms, Organic Waste and Environmental Management, C.A. Edwards; N.Q. Arancon; R.L. Sherman, (Eds.), 1-14, CRC Press, ISBN 9781439809877

[2] Lavelle, P., & Pashanasi, B. (1989). Soil macrofauna and land management in Peruvian Amazonia (Yurimaguas, Loreto). Pedobiologia, 33, 283-291.

[3] Maboeta M S , Rensburg L.V.: Vermicomposting of industrially produced woodchips and sewage sludge utilizing Eisenia fetida, Ecotoxicol Environ Safety; 2003 Oct;56(2):265-270. doi: 10.1016/s0147-6513(02)00101-x.

[4] Mathur, U. B., Verma, L. K., & Srivastava, J. N. (2006). Effects of vermicomposting on microbiological flora of infected biomedical waste. Journal of ISHWM, 5(I), 21-26.

[5] Edwards, C.A. and Burrows, I. (1988) The potential of earthworms composts as plant growth media. In: Edward, C.A. and Neuhauser, E.F. Eds., 'Earthworms in Waste and Environmental Management', SPB Academic Publishing, The Hague, 2132.

[6] Singh and S. Sharma, "Composting of a crop residue through treatment with microorganisms and subsequent vermicomposting," Bioresource Technology, vol. 85, no. 2, pp. 107– 111, 2002.

[7] Arancon NQ, Edwards CA, Lee SS, Yardim E. Management of plant-parasitic nematode populations by use of vermicomposts. Proceedings of Brighton Crop Protection Conference Pests and Diseases. 2002;8(B2):705-716

[8] Edwards, C.A. and Bohlen, P.J. (1996) Biology and Ecology of Earthworms. 3rd Edition, Chapman & Hall, London

[9] Hand, P., Hayes, W.A., Frankland, J.C., Satchell, J.E., 1988. The vermicomposting of cow slurry. Pedobiologia 31, 199±209

[10] Atiyeh, R.M., Lee, S.S., Edwards, C.A., Arancon, N.Q., Metzger, J. (2002) The influence of humic acid derived from earthworm-processed organic waste on plant growth. Bioresource Technology 84, 7-14.

[11] Dash, MC. and B.K. Senapati 1985. Vermitechnology: potentiality of India.i earthworms for Vermicomposting and vermifeed. proc. Soil. Bio. Symp. Hisar, pp.61 - 69.

[12] Hartenstein R (1983) Assimilation by earthworm Eisenia fetida. In: Satchell JE (ed) Earthworm ecology. From Darwin to vermiculture. Chapman and Hall, London, pp. 297-308

[13] Mitchell A, Edwards CA (1997) The production of vermicompost using Eisenia fetida from cattle manure. Soil Biol Biochem 29:3-4

[14] Rengel Z. 2001. Genotypic differences in micronutrient use efficiency in crops. Communications in Soil Science and Plant Analysis 32: 1163-1186

[15] Hobbelen PH, Koolhaas JE, van Gestel CA. Bioaccumulation of heavy metals in the earthworms Lumbricus rubellus and Aporrectodea caliginosa in relation to total and available metal concentrations in field soils. Environ Pollut. 2006;144(2):639-646. doi:10.1016/j.envpol.2006.01.019

[16] Gutiérrez-Miceli, F.A., Santiago-Borraz, J., Montes Molina, J.A., Nafate, C.C., Abdud- Archila, M., Oliva Llaven, M.A., Rincón-Rosales, R. and

Deendoven L. (2007). Vermicompost as a soil supplement to improve growth, yield and fruit quality of tomato (Lycopersicum esculentum). Bioresource Technology 98, 2781-2786.

[17] Gajalakshmi,S. and Abbasi, S.A. (2004). Neem leaves as a source of fertilizer-cum-pesticide vermicompost. Bioresource Technology 92, 291-296.

[18] Arancon, N.Q., Edwards, C.A., Bierman, P., Metzger, J.D. and Lucht, C. (2005). Effects of vermicomposts produced from cattle manure, food waste and paper waste on the growth and yield of peppers in the field. Pedobiologia, 49, 297-306.

[19] Lazcano, C., Revilla P., Malvar, R.A. and Domínguez, J. (2011). Yield and fruit quality of four sweet corn hybrids (Zea mays) under conventional and integrated fertilization with vermicompost. Journal of the Science of Food and Agriculture.

[20] Prabha, M.L., Jayraay, I.A., Jayraay, R. and Rao, D.S. (2007). Effect of vermicompost on growth parameters of selected vegetable and medicinal plants. Asian Journal of microbiology, Biotechnology and Environmental Sciences, 9(2), 321-326.

[21] Bhattacharjee, G., Chaudhuri, P.S. and Datta, M. (2001). Response of paddy (Var. TRC-87- 251) crop on amendment of the field with different levels of vermicompost. Asian Journal of Microbiology, Biotechnology and Environmental Sciences, 3 (3), 191-196.

[22] Chand, S., Pande, P., Prasad, A., Anwar, M. and Patra, D.D. (2007). Influence of integrated supply of vermicompost and zinc-enriched compost with two graded levels of iron and zinc on the productivity of geranium. Communications in Soil Science and Plant Analysis, 38, 2581-2599.

[23] Donald, D.G.M. and Visser, L.B. (1989). Vermicompost as a possible growth medium for the production of commercial forest nursery stock. Appl. Plant Sci. 3, 110-113.

[24] Tejada, M., Gonzalez, J.L., Hernandez, M.T. and Garcia, C., (2008). Agricultural use of leachates obtained from two different vermicomposting processes, Bioresource Technology, 99, 6228-6232.

[25] Karmegam, N., Alagumalai, K. and Daniel, T. (1999). Effect of vermicompost on the growth and yield of green gram (*Phaseolus aureus* Roxb.). Tropical Agriculture 76, 143-146.

[26] Zaller, J.G. (2007). Vermicompost as a substitute for peat in potting media: Effects on germination, biomass allocation, yields and fruit quality of three tomato varieties. Scientia Horticulturae, 112, 191-199

[27] Edwards, C.A., Arancon, N.Q. and Greytak, S. (2006). Effects of vermicompost teas on plant growth and disease. BioCycle 47, 28-31.

[28] Lazcano, C., Arnold, J., Tato, A., Zaller, J.G. and Domínguez, J. (2009). Compost and vermicompost as nursery pot components: Effects on tomato plant growth and morphology. Spanish Journal of Agricultural Research 7, 944-951.

[29] Arancon, N.Q., Edwards, C.A., Babenko, A., Cannon, J., Galvis, P. and Metzger, J.D. (2008). Influences of vermicomposts, produced by earthworms and microorganisms from cattle manure, food waste and paper waste, on the germination, growth and flowering of petunias in the greenhouse, Applied Soil Ecology 39, 91-99.

[30] Singh, R., Sharma, R.R., Kumar, S., Gupta, R.K. and Patil, R.T. (2008). Vermicompost substitution influences growth, physiological disorders, fruit

yield and quality of strawberry (Fragaria x ananassa Duch.). Bioresource Technology, 99, 8507-8511.

[31] Haydock, P. P. J., Woods, S. R., Grove, I. G., and Hare, M. C. (2013). "Chemical control of nematodes," in Plant Nematology, eds R. N. Perry and M. Moens (Wallingord: CABI), 259-279.

[32] Gabour EI, Marahatta SP, Lau J-W. Vermicomposting: A potential management approach for the reniform nematode, Rotylenchulus reniformis. Nematropica. 2015;45(1):285-287

[33] Edwards CA, Arancon NQ, Emerson E, Pulliam R. Suppression of plant-parasitic nematodes and arthropod pests by vermicompost teas. Biocycle. 2007;48(12):1-6

[34] Rodríguez-Kábana R. Organic and inorganic nitrogen amendments to soil as nematode suppressants. Journal of Nematology. 1986;18(2):129-135

[35] Kerry B. Fungal parasites of cysts nematodes. In: Edwards CA, Stinner BR, Stinner D, Rabatin S, editors. Biological Interaction in Soils. Amsterdam: Elsevier; 1998. pp. 293-306

[36] Siddiqui ZA, Mahmood I. Role of bacteria in the management of plant-parasitic nematodes: A review. Bioresource Technology. 1999;69(2):167-179

[37] Bilgrami L. Evaluation of the predation abilities of the mite *Hypoaspis calcuttaensis*, predaceous on plant and soil nematodes. Fundamental & Applied Nematology. 1997;20:96-97

[38] Sikora, R.A. & Fernandez, E. 2005. Nematode parasites of vegetables. In: Luc, M., Sikora, R.A. & Bridge, J. (Eds). Plant-parasitic nematodes in subtropical and tropical agriculture. CABI Publishing, Wallingford, UK, 319-392 pp.

[39] Akhtar, M. & Malik, A. 2000. Roles of organic soil amendments and soil organisms in the biological control of plant-parasitic nematodes: A review. Bioresource Technology 74, 35-47.

[40] Agyarko, K. & Asante, J.S. 2005. Nematode dynamics in soil amended with neem leaves and poultry manure. Asian Journal of Plant Sciences 4, 426-428.

[41] Oka, Y., Nacar, S., Putievsky, E., Ravid, U., Yaniv, Z. & Spiegel, Y. 2000. Nematicidal activity of essential oils and their components against the root-knot nematode. Phytopathology 90, 710-715.

[42] Ashraf, M.S. & Khan, T.A. 2007. Efficacy of *Gliocladium virens* and *Talaromyces flavus* with and without organic amendments against Meloidogyne javanica infecting eggplant. Asian Journal of Plant Pathology 1, 18-21.

[43] Radwan, M.A., Abu-Elamayem, M.M., Kassem, M.I. & El-Maadawy, E.K. 2004. Management of Meloidogyne incognita rootknot nematode by integration with either organic amendments or carbofuran. Pakistan Journal of Nematology 22, 135-142.

[44] Pokharel RR and HJ Larsen. "The importance and management of phytoparasitic nematodes in western Colorado fruit orchards". Journal of nematology 39 (2007): 96.

[45] Pokharel RR., et al. "Plant-parasitic nematodes, soil and root health in Colorado onion fields". In: Godin, R. (ed.). Western Colorado Research Center, Colorado State University. Annual report (2009): 39-44.

[46] Frank S Hay. The American Phytopathological Society (APS). Nematodes the good, the bad and the ugly. University of Tasmania (2019).

[47] Koenning S., et al. "Survey of crop losses in response to phytoparasitic nematodes in the United States for 1994". Journal of Nematology 31 (1999): 587-618.

[48] Renčo M., Sasanelli N., D'Addabbo T., Papajová I. 2010. Soil nematode community changes associated with composts amendment. Nematology 12 (5): 681-692.

[49] Bilgrami A.L. 1996. Evaluation of the predation abilities of the mite *Hypoaspis calcuttaensis*, predaceous on plant and soil nematodes. Fundamental & Applied Nematology 20: 96-98

Chapter 4

Plant Parasitic Nematodes: A Major Constraint in Fruit Production

Nishi Keshari and Gurram Mallikarjun

Abstract

The plant parasitic nematodes are one of the major limiting factors in fruit trees specially in citrus, banana, papaya, jackfruit, guava etc. The root knot nematodes are the major problem amongst all those nematodes infecting on these trees. Besides, directly causing a huge losses, they are also inviting the secondary plant pathogens, like fungi, bacteria, viruses etc. amongst which, the wilt fungus, *Fusarium* species increase the severity of the diseases. This complex disease is becoming much severe in banana and guava recent years. In citrus also, the citrus nematodes, *Tylenchulus semipenetrans*, is causing havoc by slow decline disease and it is becoming a major problem in horticultural nurseries because these nurseries are a hot spot of citrus nematodes. So, unknowingly these nematodes get spread to different places. The management of these nematodes by simple, cheap and eco friendly methods, is very important as it will decrease the monetary pressure on cultivators as well as it helps in improving environmental pollution.

Keywords: plant parasitic nematodes, fruits, *Meloidogyne* spp., *Tylenchulus semipenetrans*, *Pratylenchus* spp.

1. Introduction

Plant parasitic nematodes cause considerable economic losses in fruit crops. The main loss is the destruction of roots which hinders the movement of nutrients and water through the vascular system, so, there is drastic reduction in fruit or bunch weights, the quality of fruits is deteriorated and there is a drastic reduction in plant numbers. Furthermore, roots damaged by nematodes are easy prey to fungi and bacteria which invade the infected roots and feed on them and thus roots decay rapidly. The root-knot nematode, *Meloidogyne incognita*; the burrowing nematode, *Radopholus similis* and citrus nematode, *Tylenchulus semipenetrans* are the major nematode pests that infect these fruit crops. Around 30–40% loss in yield is due to these nematodes. The nematode infestation in fruit crops not only aggravate disease complexes but also breaks down disease resistance in certain varieties of fruit crops. These nematodes mainly spread through infested planting material to other uninfested sites. For an instance, in banana, paring or trimming of suckers is carried out before planting which is usually not adequate to eliminate deep infections in the suckers. The residual nematode population builds up and disseminates when they enter the irrigation system. Other routes of their dispersal include soil adhering to tractor tyres, shoes of labour and tillage implements. The major fruit crops which suffer severe nematode infestation, are discussed here.

2. Citrus

Citrus is grown in more than 125 countries in a belt between 35° latitude north or south of the equator. The citrus is generally consumed as fresh fruit, approximately 68% of the world's citrus production, and in international trade, about 11% of total production is used [1]. *Citrus* spp. are naturally deep-rooted plants [2, 3], and optimum growth requires deep, well-drained soils. The first nematode discovered in citrus, was, *Meloidogyne* sp. which parasitized the citrus in Florida in 1889 but people were unaware about these nematodes. Again in 1913, Thomas discovered the citrus nematode infecting citrus in California. Nathan Cobb had reported it as a new species, *T. semipenetrans*, and this was the causal agent of mottling disease in citrus in California, later identified as 'slow decline' because the trees declined in vigour but very slowly about in 10–15 years. *T. semipenetrans* has been found in every citrus-growing region of the world since its discovery [4]. Field infestations within United States, infect 50–90% of citrus orchards in Arizona, California, Florida, and Texas, as well as local vineyards in California [4–6]. The major economic nematodes causing diseases in citrus, are *T. semipenetrans*, *Pratylenchus* spp. and *Meloidogyne* spp.

2.1 Citrus nematode (*T. semipenetrans*)

In world, Cobb in [7] first described its distribution, morphology and life history. In India, it was first reported by Siddiqi [8] at Aligarh, Uttar Pradesh. About 80% of the citrus trees were reported to be infected with this nematode. This slow decline further results in 'die back' disease in most of the citrus trees in India and is a major problem that is estimated at 8.7–12.2%. The citrus nematodes are found highly in number when the citrus orchards are established in the soil which is finely-textured or sandy having high organic matter content. Nematode reproduction was positively favoured when there are fluctuations in soil salinity from high to low, while sandy soils poor in organic matter, hinder population increase [9]. In Florida and California (USA), there is 24–60% of *T. semipenetrans* infection in citrus orchards while it is 70–90% in commercial orchards of Brazil and Spain [10]. This shows that the citrus nematodes can infect under extreme range of environmental conditions. Unfortunately, infected nursery seedlings which are the main infection material transported from one to another, are not easily detected by the personnel involved in that business just because of unawareness [10]. If the infection is not severe, the roots show only lesions but, during severe infection, the sloughing of root cortex appears and roots die finally. Nematode infection increases the levels of the cell-damaging enzymes [11].

2.2 Symptoms

1. Reduction in vigour of the plant

2. Reduction in tree growth

3. Reduction in fruit number

4. Reduction in fruit size

5. Decline is from upper side to the lower side of the plant, so, called 'die back'

6. If the initial population is high, the death of plants occurs in very early age

7. The roots became sticky because of gelatinous matrix secreted by the females and soil particles glued to the roots which do not go after washing with water

8. The cortex separates easily if there is heavy infection because the females are feeding semiendoparasitically on the roots having their posterior two third of body part outside the root where they lay eggs and excrete the gelatinous matrix to protect the eggs

9. Leaves become smaller and chlorotic

10. Leaves fall due to poor vigour

11. If there is water stress during infestation, then wilting occurs

12. There is poor root development

13. The feeder roots decay very fast

14. Root death occurs due to heavy infestation

15. The diseased trees become dwarf and yield less than the healthy ones

16. Duncan [4] reported that the symptoms like reduced leaf and fruit size, canopy thinning, and die back of upper branches, are the most conspicuous symptoms of slow decline and that result in less yield

2.3 Life cycle

T. semipenetrans exhibits sexual dimorphism, i.e., different shape of male and female individuals at both the juvenile and adult stage. The life cycle duration is of 6–8 weeks from egg to egg [12]. The *T. semipenetrans* biology and ecology have been extensively studied [10]. The egg hatching takes place in 12–14 days at 24°C. The male larvae after second stage, do not feed and become mature in 7 days whereas the female takes 14 days to find the feeding site on the root and start feeding and moulting. The female juveniles can survive more than 2 years in the absence of roots [13]. The life cycle was of 14 weeks on *Poncirus trifoliate*, 10 weeks on *Ruta bracteosa* and 7 weeks on *Citrus aurantianum* and *C. limettoides* [14]. The mature males are vermiform and mobile found in the soil or in the egg masses. Therefore, the feeding apparatus (stylet and oesophagus) of adult males is poorly developed and may be difficult to observe. *T. semipenetrans* is a sexually reproducing species that can occasionally reproduce by facultative parthenogenesis without the need of males. The mature females and their eggs are found attached to roots which are protected by soil particles that sticks to gelatinous matrix. The females are swollen and enlarged posteriorly often protruding on the root surface in a finger like protrusions while elongated and not swollen anteriorly generally embedded and hidden in cortical parenchyma. After hatching at optimum temperature, i.e., at 25°C, females lay eggs after 6 weeks, on the root surface in a gelatinous egg mass secreted from the excretory pore.

2.4 Histopathology

The second stage larvae enter the root surface and start feeding on the mature part. After moulting, the immature females penetrate deeper in the cortex region and their neck becomes longer to feed inside. The posterior portion remains outside of the root. They establish a feeding site around their stylet where the cortical cells change into food sink by reaction of dorsal oesophageal gland secretions. These are called 'nurse cells' which provide food to the developing females. The nurse cells are thick-walled cells with modified cell organelles like enlarged nucleus and nucleolus. These cells have no vacuoles. The cells are gradually destroyed by their feeding and hence the plants can not draw food and water for their growth, so change in development proceeds which finally results in poor vigour.

2.5 Host range

Unlike many nematodes, *T. semipenetrans* has a restricted host range. Many plants belonging to Rutaceae family, were found as hosts of this nematode. It was reported that from 23 countries, 29 species of *Citrus*, 21 citrus hybrids and 11 other species as the hosts [15]. Except these plants, it was also reported to attack on other plants like *Andropogon rhizomatus*, *Panicum* spp., Olive (*Olea* spp.), grapevines (*Vitis* spp.), Persimmon (*Diospyrous lotus*), Pear (*Pyrus communis*), *Calodendrum capense*, climbing hemp weed (*Mikania batatifolia*) and Lilac (*Syringa vulgaris*) [16–19]. Parvatha Reddy and Singh [20] reported that the citrus nematode also attacked grapes and loquats. It can parasitize more than 75 plant species belonging to rutaceous species (especially citrus and their close relatives) which are its suitable hosts [13]. Till now, there have been no reports of *T. semipenetrans* infecting herbaceous plants [17]. El-Mohamedy et al. [21] reported numerous citrus varieties from Egypt, Washington Navel, Valencia orange, Mandarins group varieties (*C. reticulata*), lemon (*C. aurantifolia*), and Balady orange (*C. sinensis*), Grapefruit (*C.* × *paradisi*), Sour orange (*C. aurantium*), and Kumquat (*C. japonica*) infected with this nematode [22, 23].

2.6 Complex disease

The citrus nematode also interacts with other plant pathogens and increase the severity of the disease. The wilt fungus, *Fusarium oxysporum* and *F. solani* with citrus nematode caused the death of citrus trees. The interaction between *T. semipenetrans* with such microorganisms occurs in inconsistent ways. It can reduce the infection of roots by *Phytophthora nicotianae* after the infection to citrus seedlings and it can also increase the virulence of *Fusarium solani* [10]. It was reported that the high population level of nematodes when interacted with *F. semitectum*, got synergistic effect on the infected citrus seedlings [24]. *Fusarium* spp. can be pathogenic on citrus roots alone [25] or in combination with nematodes [26], which leads to the great destruction of the feeder roots. The loss of feeder roots due to feeding of nematodes results in increase of drought stress and decrease of soil nutrient uptake, leading to chlorosis and loss of leaves. Affected trees do not die, but have an unthrifty appearance and yield fewer, smaller fruits than uninfested trees.

2.7 Nematode spread

The major cause for the spread of this nematode is because of distribution of infested planting material from the horticultural nurseries. Once the infested soil is taken with the planting material to distant places, it will spread this nematode to new sites. The other spreading agents are human and animals with the infested soil on their feet, agricultural implements, and water. They can survive in the soil for long periods in the absence of host that enables them to infect after a long time also. The main source of infection in the citrus plant, are, infected seedlings, organic fertilizers, plant materials, irrigation, and machinery which are affecting growth and yield in the newly planted area [27]. In Egypt, which is a highly ranked citrus producing country [27], the citrus orchards were incorporated with the soil brought from silty soil from the Nile Valley for mulching and improving the soil quality but that soil was nematode infested, so the disease incidence got aggravated [28]. So, with time, the nematode spread their populations in that soil and the losses increased [27]. Soil moisture is often inversely related to population growth of *T. semipenetrans* [29–31].

2.8 Management

2.8.1 Preventive measures

The most important and effective method is to take every effort to avoid the use of infested planting materials and contaminated farm implements when new plantations have to be established. In the orchard, proper drainage and light should be there and shade should be avoided as far as possible. New nurseries should not be established near the old citrus orchards. All sanitation practices should be taken to avoid nematode infestations. Use of certified nematode-free material for planting, is also very important. If there is established infection, the citrus orchard should be rotated with annual crops for 1–3 years before replanting helps to reduce citrus nematode populations. For intercropping, Marigold is an excellent crop which has repellent action and reduces the population of nematodes in citrus [32].

2.8.2 Biological measures

Application of *Pseudomonas fluorescens* @ 20 g/tree. *Paecilomyces lilacinus* parasitize nematode eggs and females, reduces the number of plant parasitic nematodes in soil, *T. semipenetrans*. Park et al. [33] reported that *P. lilacinus* could produce leucino toxin and other nematicidal compounds. *Trichoderma* spp. play major roles in controlling plant diseases in roots and soil. The *Trichoderma* spp. have antagonistic activities to be used as effective biological control agents for many plant diseases which are caused by soil borne fungi and nematodes [34]. Although *Bacillus subtilis* was reported as a bio-agent against soil borne fungi [35] some strains of *B. subtilis* exhibited enormous potential as bioagent in the management of nematodes [36]. *B. subtilis* produces antibiotics as bacterocin and subitisin [37, 38]. Streptomycetes are the major group of actinomycetes producing secondary metabolites that could decrease the invasive juveniles of root-knot nematodes. Streptomyces is known for its chitinolytic activity which produces more extracellular chitinase [39]. Some species of streptomycetes release compounds like antibiotic that inhibit the growth of plant-pathogenic fungi [40] and plant parasitic nematodes [41]. Qingfei et al. stated that *Streptomyces* spp. produce lytic enzymes and nematicidal compounds and can be one of candidates for bio-agents against nematodes. Le Roux et al. [42] demonstrated that *P. lilacinum* individually controlled *T. semipenetrans* Cobb on mandarin and rough lemon effectively, but when the fungus was combined with oil-cakes, the results were more significant.

2.8.3 Chemical measures

In heavy infestation, many nematicides have successfully been used to lower down the population of *T. semipenetrans* on citrus in many locations. The soil treatments with soil solarization and nematicides is highly beneficial in both replanting conditions and already established orchards. Pre-plant application of carbofuran 3 G @

100 g/tree, was found highly beneficial. The nematicide application often increases the citrus yield [10]. Nemastop (natural oils) as commercial nematicide, play very important role in controlling nematodes. The effect of Nemastop on the nematode might be due to alkyl cysteine sulphoxides which released a mixture of volatile alkyl thiols and sulphides [43]. Whereas, Nemaphos belonging to organophosphate group [44] showed a highly performance systemic nematicide. When halogenated hydrocarbons are used as pre-plant soil fumigants, these can effectively control *T. semipenetrans* for many years [45–48]. However, to maintain the low population and higher yield, one has to apply the chemicals repeatedly. In the first year of treatment, the effect will be little on yield and fruit quality but the efficacy to increase the growth and yield parameters can be observed in the following year [49–51].

Oxime carbamates (aldicarb, oxamyl, Carbofuran) and organophosphates (fenamiphos, ethoprophos and cadusaphos) are the two main groups of nematicides which are available in the market for the management of citrus nematode. Of these, granular formulation of Cadusaphos has shown greater efficacy against the *T. semipenetrans* [52–55]. Irrigation is generally recommended before nematicide application for better results.

2.8.4 Resistant rootstocks

The use of resistant root stock is the best method to avoid the disease if available. It was reported by many workers that the use of resistant (Swingle citrumelo) rootstocks and certified propagative material which are free from nematode parasites of citrus, are promising cultivars for preventing the loss caused by *T. semipenetrans* to citrus [56–58]. In Florida, this approach has significantly reduced the spread of this parasite, making the land free from nematode infestations [59]. In California vineyards, resistant (Ramsey) or moderately resistant (vinfera Dog Ridge) grape rootstocks were used successfully [60] (McKenry, personal communication). To get sustainable agriculture, planting of nematode certified citrus and grape rootstocks, is an excellent practice that should be adopted for other fruit crops also which are susceptible to nematode infections. Resistance-breaking biotypes were developed on Swingle citrumelo [61]. The commercially resistant rootstock, Swingle citrumelo is common in Florida and combined with regulation program of the citrus nematode, has decreased the spread of *T. semipenetrans* dramatically [62]. Using a resistant rootstock is recommended whether or not nematodes are present. Trifoliate orange is known to be tolerant to citrus nematode.

2.8.5 Soil solarization

Soil solarization is an effective method to disinfest the upper soil layers by moistening the soil and covering with a clear plastic sheet in regions with hot and dry summer months. This method is highly beneficial to manage the population of insects, soil borne pathogens, weed seeds and nematodes by altering the physical, chemical, and biological properties of the soil. In South Africa, solarization has not shown promising and there was inconsistent suppression of the citrus nematode and tree growth [63], which may be because these nematodes are found deep within the soil profile and so, are not affected by solarization that is most effective for the upper soil layers [64].

2.8.6 Steam treatment

Steam treatment of soil is widely used for the control of nematodes in planting material and is shown very effective. In this method, the soil is heated up to 70°C, mainly by means of aerated steam. It is very useful and economical for disinfestation of nursery beds. Steam treatment of vermiculite or tuff stones is usually effective but is more difficult for peat soils due to their high water content [65]. The dipping of planting material in hot water is also effective but here the temperature of water should be taken care of, otherwise the germination may get affected. Bare root dipping of citrus seedling in hot water at 45°C for 25 min [66] or 50°C for 10–20 min [67] was found to be effective without any adverse effect on the germination.

2.9 Lesion nematodes (*Pratylenchus* spp.)

There are three species of *Pratylenchus* which can affect citrus i.e., *P. coffeae*, *P. brachyurus*, and *P. vulnus*. All are reported from Egyptian citrus orchards [68]. The most pathogenic species is *P. coffeae* [10]. Lesion nematodes, being a migratory endoparasites, cause infection mainly in the feeder roots during their movement by penetrating the cortical tissue, but they do not invade the vascular tissues. After their infection, the secondary organisms infect the root tissues and then the vascular tissues also got infected. *P. coffeae* is obligatorily amphimictic, all stages infective with males feeding in the roots [69]. Its reproduction is highest at high (26–30°C) soil temperatures. At those temperatures, the life cycle is completed in less than a month, and it can achieve densities of up to 10,000 nematodes/g root [70] and persist in soil roots for at least 4 months. This leads to root weight reduction by half and growth reduction ranging from 49 to 80% in young trees in field conditions. A 3-fold to 20-fold differences between infected and non-infected trees was observed in terms of the numbers of fruit [71]. Commercial rootstocks resistant to *P. coffeae* are yet to be identified. A lesion nematode, *P. coffeae*, was detected on citrus in Sao Paulo State, Brazil and found to infest about 1% of the nurseries and orchards [72].

The biology of *P. brachyurus* is similar to that of *P. coffeae.* [13]. It has been established as a pathogen of citrus seedlings across several soil types. [73]. After controlling *P. brachyurus* with aldicarb, yields of Valencia orange trees grafted on rough lemon were increased, and plants sustained reduced frost damage in the winter [51]. Some studies failed to note the fact that *P. vulnus* has been found associated with citrus plants in Egypt [10, 13, 68]. This species is capable of causing significant damage to citrus seedlings but has not been reported to damage mature plants [74]. Biology, population growth rates, and root damage are similar to those described for *P. coffeae* [13]. Several *Pratylenchus* species have been identified in Egyptian citrus orchards based on field studies [68].

Host range: *Citrus limon*, *C. sinensis*, *C. reticulata*, banana [75] and *Citrus jambhiri* [76].

2.9.1 Management

Chemical: Fensulfothion or phenamiphos @ 4.4 kg a. i./ha and aldicarb or carbofuran @ 4 kg a. i./ha [77].

Resistant root stocks: Trifoliate orange (*Poncirus trifoliata*), Rubiboux 70-A5, hybrids of *Microcitrus australis* x *M. australasica.*

*2.9.2 Root knot nematodes (*Meloidogyne *spp.)*

Root knot nematodes attacking citrus trees, are not reported much and is little in distribution [10]. Only a few locations in world, have been found to be infested with this nematode. A pathogenic root knot nematode species (known as Asiatic pyroid citrus nematode) recorded from Taiwan and New Delhi could cause elongated galls on citrus roots [13]. It can complete its life cycle on several citrus and other plant species. The common species are *Meloidogyne incognita*, *M. javanica*, and *M.*

arenaria reported to infect roots of Troyer citrange and sour orange, causing small galls, but their multiplication is not recorded [78]. *M. indica* was reported from citrus tree at some locations in India [79].

2.9.3 Symptoms

1. Severe stunting of the plants
2. Yellowing of the leaves
3. Plants show unthriftiness
4. Symptoms of twig blight appear
5. Slow drying of the tree
6. Small to large galls on the roots are the characteristic symptoms
7. Poorly developed root system [80]
8. The trees do not flower and fruit
9. Large cavities are found on the roots
10. Egg masses float on the root surface

2.9.4 Management

Chemical methods: Seedling dip treatment with carbosulfan @ 500 ppm for 6 h can effectively control root knot nematodes.

Organic amendments: Mustard cake, farm yard manure and poultry manure @ 2.5 kg/plant were found effective against root knot nematode and increasing the plant growth.

Host resistance: Resistant rootstocks, like, Rangpur lime 8784, Sour orange Tirupati, Citrumello 4479, Rangpur 8748, Rangpur lime chethalli, Trifoliate orange chethalli, Nasnaran, Hazara Australia, Rangpur lime Kirumakki, Pramalini and Anand Selection were moderately resistant.

3. Guava

The common guava (*Psidium guajava* L.) is indigenous to tropical America. It is a popular fruit generally consumed as fresh fruit but also processed as jam, paste, puree, canned shells and juice for round the year use. It is grown throughout the tropics and subtropics and is of commercial importance in more than 60 countries [81]. This fruit tree also suffers many nematode infections. Mostly root knot nematodes were reported from guava roots and after infection, the roots are predisposed to wilt fungus which is common wilt disease causing pathogen found in guava. This interaction resulted increased disease severity. Khan et al. [82] observed greater damage to guava with both *Helicotylenchus dihystera* and *Fusarium oxysporum* than with the nematode alone. In India, *Hoplolaimus indicus* was found to be a pathogen of guava [83, 84] and *Tylenchorhynchus cylindricus*, in numbers of up to 2000 nematodes/100 cm^3 of soil, was found associated with damaged guava trees in Iran [85]. In the past two decades, the root-knot nematodes, *Meloidogyne* spp. Göldi, have been reported on some species of the tropical fruit trees grown in the region [86]. Gomes et al. [87] demonstrated that guava trees infected with *M. mayaguensis* had deficiencies in nitrogen, potassium, phosphorus, calcium, and magnesium, and that these mineral deficiencies were proportionally related to the severity of root galling and root decay, which eventually led to the death of guava trees within a few months. Recently a new species, *M. enterolobii* has been found to be widely associated with many guava trees.

3.1 Symptoms

1. Yellowing of trees
2. Leaves shed prematurely
3. The branches start drying
4. The trees show less vigour
5. The trees fruit very less
6. The affected trees dry within three months
7. Multiple small galls can be seen on the roots
8. The affected roots show necrosis
9. There is decrease of feeder roots [88]

3.2 Spread

In guava, grafts are often produced in polythene bags with a substrate mixture (sand + soil + FYM or other organic manure). In most cases the substrate mixtures harbour the aforementioned nematodes as well as other harmful fungi and bacteria. Generally, nurserymen do not treat the soil combination used in the production of fruit seedlings or grafts in their nurseries. As a reason, before applying the substrate, it should be treated with biopesticides.

3.3 Management

3.3.1 Prevention

Nematode free plants should be used for new planting and the orchard should be planted in nematode free soil. The soil used for the preparation of new plants should be sterilized. The equipment should be sterilised before using.

3.3.2 Deep summer ploughing

The soil for the planting of new guava plants, should be deeply ploughed during hot summer months. It should be repeated twice or thrice at 15 days interval.

3.3.3 Biological method

For biological control, many fungal and bacterial bioagents are used for the management of nematodes. The fungal bioagents, *Purpureocillium lilacinum* and

Verticillium clamydosporium were found as the most potent fungal parasites that can effectively control *Meloidogyne* spp. on many host plants [89]. Another fungal bioagent, *Trichoderma harzianum* effectively suppressed the population of the root-knot nematode, *M. enterolobii* in both soil and roots of guava in Thailand [90]. In Saudi Arabia, Al-Hazmi et al. [91] found a heavy colonization of the cysts of *Heterodera avenae* Woll. with the fungus, *Verticillium clamydosporium.* Rao [92] who reported that the fungus, *P. lilacinum* and the bacteria, *P. fluorescens* enriched the farm yard manure, fairly controlled the reniform nematode, *R. reniformis* and the root knot nematodes.

3.3.4 Organic and inorganic nitrogenous amendments

These amendments are useful both for managing plant parasitic nematodes as well as to improve the soil fertility [93]. At rates as low as 300–400 mg/kg soil, urea and ammonia were found to be effective in controlling plant parasitic nematodes [94]. Guava decline disease, a disease complex caused by *M. mayaguensis* in association with the fungus, *Fusarium solani*, has been successfully managed in a commercial guava plantations, with significant yield gains obtained by the use of cow manure and poultry compost [95]. Urea and nitrogenous fertilizer are considered to be good nematicides when applied at levels as low as 300–400 mg/kg soil because due to root infections, the roots become unable to take the minerals and nutrients [94, 96–98]. Previously it was also reported that organic and inorganic nitrogen amendments had a nematicidal effect against plant parasitic nematodes [93, 94, 99]. Gomes et al. [95] reported that the guava canopy when treated with organic soil amendments, particularly poultry compost and cow manure, gave better control of the root knot nematode, *M. mayaguensis*. Organic soil amendments not only promote the growth of soil microorganisms that are antagonistic to plant parasitic nematodes but also release specific toxic compounds during their decomposition that may have nematicidal effects against nematodes [99]. These organic amendments also improve crop nutrition and growth which lead to increased tolerance of plants against plant-parasitic nematodes [100].

3.4 Production of healthy rootstocks/grafts

Enriched substrates should be used for producing healthy grafts or rootstocks of fruit crops. A mixture of enriched FYM can be produced by mixing one ton of soil mixture consisting red soil and sand in equal proportions with 500 kg of FYM which is added with 2 kg each of *P. fluorescens* 1% W. P., *Trichoderma harzianum* 1% W. P., *Paecilomyces lilacinus* 1% W.P. + 50 kg neem or pongamia cake +5 kg carbofuran or phorate.

3.5 Spraying or drenching the nursery seedlings or grafts with bio-pesticides

The nursery seedlings or grafts can be treated by dissolving 5 g or 5 ml/l of water once in 10 days.

3.6 Management of nematodes in the main field

3.6.1 Soil application

Before planting the saplings of guava, the land should be thoroughly ploughed and soil should be brought to fine tilth. Then recommended doses of fertilisers should be added because the vigour plants resist the nematode attack. Now carbofuran or phorate @ 20 kg–25 kg + 200 g neem/pongamia/mahua cake per acre should be added to reduce the initial population of nematodes. The optimum moisture should be maintained in the beds which is necessary for proper decomposition of organic matter. In case of organic farming, during the land preparation, application of two tons of FYM or 500 kg of neem cake/pongamia cake or one ton of enriched vermicompost with *P. fluorescens* + *Trichoderma harzianum* + *Paecilomyces lilacinus*.

3.6.2 Process of enrichment of FYM

For enrichment of FYM, 2 kg each of *P. fluorescens, Trichoderma harzianum* and *Paecilomyces lilacinus* formulation should be mixed with one ton of well decomposed FYM under shade and covered with mulch. An optimum of 25–30% moisture should be maintained for a period of 15 days. This mixture should be thoroughly mixed once in a week to promote maximum multiplication along with even growth of the microorganisms in the entire lot of FYM.

3.6.3 Process of enrichment of neem cake

Neem cake can be enriched by mixing with 2 kg each of *P. fluorescens*, *Trichoderma harzianum* and *Paecilomyces lilacinus* with one ton of neem cake under shade and covered with mulch. For next 15 days, an optimum moisture of 25–30% has to be maintained followed by thorough mixing once in a week to ensure maximum multiplication & uniform growth of the microorganisms in the entire lot of neem cake.

3.6.4 Process of enrichment of vermicompost

Similarly, vermicompost can be enriched by mixing with 2 kg each of *P. fluorescens*, *Trichoderma harzianum* and *Paecilomyces lilacinus* with one ton of vermicompost under shade and covered with mulch. For next 15 days, an optimum moisture of 25–30% has to be maintained followed by thorough mixing once in a week to ensure maximum multiplication & even spread of the microorganisms in the entire lot of vermicompost.

3.6.5 Application of bio-pesticides at the time of planting

At the time of planting, application of bio-pesticide enriched FYM @ 3 kg or enriched neem cake @ 250 g or enriched vermicompost @ 500 g/plant at an interval of six months.

3.6.6 Spraying

Spraying plants with organic formulation containing *P. fluorescens* & *Trichoderma harzianum* at regular 30-day intervals at a dosage of 5 g/l or 5 ml/l.

3.6.7 Drenching or application through drip irrigation system

Drenching of the above biopesticide @ 5 g/l or 5 ml/l at regular interval of 30 days.

4. Banana

It is also one of the important fruit crops in India, cultivated over 0.83 million ha, constituting about 44.3% of total fruit production. Generally, the farmers and cultivators grow banana as cash crop and do the intensive cultivation on a commercial scale and as monoculture in the same field. This system invited several pests and diseases to this crop. Among these major biotic stresses, plant parasitic nematodes are causing severe losses in banana production not only individually but also with the interaction of wilt fungi, *Fusarium oxysporum f.* sp. *cubense* [101]. The major nematode causing economic yield loss globally, is the burrowing nematode, *Radopholus similis* (Cobb, 1893) Thorne, 1949. Other nematode species that are found in banana cultivation along with *R. similis*, are, lesion nematode (*Pratylenchus coffeae*), spiral nematode (*Helicotylenchus multicinctus*) and *Meloidogyne* spp. In South Africa, 34 plant parasitic nematode species have been found associated with banana [102] (SAPPNS1). Likewise, *Meloidogyne incognita* (Kofoid and White, 1919) Chitwood, 1949, *Meloidogyne javanica* (Treub, 1885) Chitwood, 1949 and *Meloidogyne arenaria* (Neal, 1889) Chitwood, 1949 are most commonly reported root knot nematode species found in association with local banana cultivars in Zimbabwe [103]. Root knot nematodes were the most abundant and together with spiral nematodes constituted 72% of the total plant parasitic nematode complex [104]. The root knot nematodes become larger in size in banana root tissue as the root tissues are thicker. These nematodes attack root and corm tissues causing damage which results in long vegetative growth cycle, late fruiting, production of small bunches, less fruiting and finally toppling of the plants. The spiral nematodes produce root lesions in the corm which attract the secondary pathogen, fungi and bacteria, that aggravate the damage of the root system [105]. Root knot nematodes are present in almost all banana plantations [106]. The general perception is that these nematode pests can cause severe damage to young plants, resulting in suboptimal growth and yield.

4.1 Host range

Both *M. incognita* and *M. javanica* has a wide host range of cultivated crops including most broad leaf weed species. To effectively control root-knot nematodes in bananas, special attention should be paid to weeding during fallowing or crop selection for rotation [107].

4.2 Losses

The diseases in banana are caused by 3–4 nematodes so, the quantification of the damage by individual species is, therefore, not possible. Willers [108] estimated that these pests caused a direct loss of 19% in total production of the crop.

4.3 Symptoms caused by burrowing nematode, *R. similis*

The origin of burrowing nematode is believed to be from Australia and New Zealand [109] from where recently new species have been described. The relative increase in worldwide distribution is mainly due to transfer of infected planting material domestically as well as internationally. The importation of banana cultivar Cavendish which is susceptible to nematode attack is often correlated with the wide distribution of *R. similis*. It is assumed that *R. similis* was introduced in Latin America and the Caribbean, on the cv. Gros Michel where it subsequently infested the more susceptible Cavendish cultivar [110]. In a study, 55 out of 57 burrowing nematode isolates, collected from Australia, Cameroon, Central America, Cuba, Dominican Republic, Florida, Guadeloupe, Hawaii, Nigeria, Honduras, Indonesia, Ivory Coast, Puerto Rico, SA and Uganda were morphologically similar to *R. similis*.

The following are the symptoms:

- Dwarfing of plants
- The size and number of leaves are reduced and bunch weight is also reduced.
- Dark red lesions appear on roots which is the identification of fungal infection after nematodes.
- The whole plant topples down in windstorms or heavy rains due to less vigorous roots after infection of plants.
- Heavily infected plants have thin pseudostems, while the leaves are yellowish or discoloured, greenish yellow bands appeared along the leaf blades [111].
- Infected plants are also prone to wilting during hot days because the plants could not take water due to root damage.

4.3.1 Management

- Field fallowing for a period of six months or longer should be done to avoid the continuity of the nematode generation.
- Crop rotation with non-host crops should be done.
- Use of disease-free planting material.
- Storage of large corms in the sun for two weeks prior to planting to kill the nematodes and their juveniles.
- Use of cover crop calapogonium is very useful to avoid the population.
- Removal of infested portions of corm before planting.
- Hot water treatment at 55°C for 15–25 min and the time and temperature limit should be strictly followed.
- Paring and Pralinage are important practice to take care the plants after infection
- Carbofuran 3 g @ 40 g/plant at 90 days after planting as this is systemic nematicide.
- Application of neem cake also enhances the plant growth as well as the management of nematodes
- Host range: Grapevine, Papaya, Pomegranate, Banana

Symptoms caused by Lesion nematode, *Pratylenchus coffeae*

- Extensive black or purple necrosis on roots, are found
- Stunting of plants as a characteristic symptom

- Reduction in size and number of leaves and in bunch weight.

Symptoms caused by Root knot nematodes, *Meloidogyne* spp. and Reniform nematode, *Rotylenchulus reniformis*

- There is not normal growth, the leaves become hard and brittle
- The fields look in patches because of abnormal growth due to distribution of nematodes.
- The yield is reduced finally.
- Heavily infested plants have a much reduced root system with large elongated galls in root-knot infested plants.
- Small galls along feeder roots and large galls on the main roots, are found in root knot infested plants.
- Rabie [112] reported that the combination of root-knot nematodes and other stress factors, such as drought and cold temperatures, are responsible in inflicting symptoms of 'False Panama Disease,' which is similar to 'Panama Disease'.
- In the former case transverse sections of rhizomes of infected plants showed reddish brown to brownish purple discoloured vascular tissue while in later case, reddish brown ring like symptoms were observed in cross section of pseudostem
- Leaf symptoms include progressive dying back of older leaves, starting at the tips. Galls occur on the primary and secondary roots, whilst distortion of roots and sometimes bifurcation occurs after heavy nematode infections.

4.3.2 Management

- Planting stocks which should be certified and free of infection.
- Application of carbofuran 3G @ 60 g/vine helps to improve the plant growth.
- Application of *P. fluorescens* @ 50 g/vine as biocontrol agent.
- Application of neem cake, vermicompost and pressmud, coriander intercrop and marigold intercrop in banana resulted in a general reduction in the population of plant parasitic nematodes. The nematicidal effect of neem cake, vermicompost, pressmud, marigold and coriander have been reported earlier [113–115].

4.3.3 Helicotylenchus *spp. (Spiral nematodes)*

Spiral nematodes are one of the most common group of nematodes infesting banana which include *H. multicinctus* (Cobb, 1893) Golden, 1956, *Helicotylenchus dihystera* (Cobb, 1893) Sher, 1961, and *Helicotylenchus erythrinae* (Zimmermann, 1904) Golden, 1956 [103, 106]. A survey of commercial banana plantations showed that species of *Helicotylenchus* (mainly *H. multicinctus*) were present in 95% of all the samples which had highest overall average [106]. These results were similar to previous studies. *H. multicinctus* was recorded in most of the soil and root samples collected in a survey [116]. Gowen and Quénéhervé [101] suggested that

H. multicinctus is often the major parasitic nematode on banana where temperature and rainfall conditions are suboptimal for the crop.

4.3.4 Damage

The above-ground symptoms are not specific and resemble damage caused by other nematode pests of banana. Symptoms of damage inflicted by *H. multicinctus* include discolouration of the root epidermis where small reddish lesions can be observed in the superficial cortical region also in rhizome tissue of infected plants along with reduction in the number of lateral roots [103]. Under severe infestations, the smaller lesions enlarge and coalesce leading to extensive necrosis in the outer cortical region of root and even root dieback sets in [117]. Toppling of plants can be observed due to poor anchorage under severe infestation [105]. No distinct biotypes or races have been reported in *H. multicinctus* [101].

4.3.5 Host range

H. multicinctus hosts range from most edible banana and plantain to various alterative host plants, such as pigweed (*Amaranthus* spp.), purslane (*Portulaca oleracea*) and ornamentals [101].

4.3.6 Nature of damage

Most damaging spiral nematode species spend most of their life cycle in root tissues of banana. Their buccal cavity is equipped with a hollow stylet which helps in puncturing and feeding the inner contents of cells. They multiply and build their population as high as a million in corm and root tissues and they alter the physical and functional integrity of the tissues. Nematodes disrupt nutrient and water uptake results in delay of growth and finally banana plants topple down. They destruct the primary roots so poor anchorage develops and that results in toppling of the plants.

4.3.7 Loss

In banana, the majority of the losses caused by *R. similis* infection is due to mass destruction of primary roots and could be present throughout entire root system, including the rhizome leading to poor anchorage [118]. During windy conditions, the plants with bunches often topple off due to heavy weight and poor anchorage, Hence the name "Toppling disease". The nematodes move laterally in cortical region and colonize the cavities caused by parasitic and saprophytic fungi which results in greater lesion formation thus, leading to indirect disruption of stele which otherwise is rare. When this occurs, the entire root beyond the initial nematode entry site becomes functionless [103].

4.3.8 Host range

Duchame and Birchfield [119] established the existence of a *R. similis* biotype that also attacked citrus, it was confirmed that the *R. similis* did not attack citrus [120, 121]. An experiment to determine the host status was conducted to study the reaction of banana plants which were planted in *R. similis* infested soil. Out of 100 plants tested, only 20 were found to be able to act as host for *R. similis* [120, 121]. However, in field conditions, no record of *R. similis* was found to be associated with banana in both the National Collection of Nematodes (NCN) and South African Plant Parasitic Nematode Survey (SAPPNS) records.

4.4 Lesion nematodes (*Pratylenchus* spp.)

The lesion nematodes which are well represented in South African region, have a limited distribution in banana plantations. Two species of lesion nematodes which are most frequently found in South Africa plantations include *Pratylenchus coffeae* (Zimmerman, 1898) Filipjev and Schuurmans Stekhoven, 1941 and *Pratylenchus brachyurus* (Godfrey, 1929) Filipjev and Schuurmans Stekhoven, 1941 with the former one being more frequently encountered [116]. *Pratylenchus coffeae* was very effective in limiting the spread of *R. similis* in local banana-growing areas as the combination of legislative measures with the development of tissue culture based propagative material for producing nematode free plants, limited the spread of *R. similis* and *P. coffeae.* Tissue culture plants are developed in the laboratory using healthy planting material are free of pests and diseases. Before these plants are distributed to the producers, they have to be tested and declared virus free which is generally found in 25.9% of root and soil samples in commercial plantations [106]. Despite, these pests are widely distributed throughout the banana-producing areas, samples from several individual plantations had no lesion nematodes. The lesion nematode population in commercial banana plantations, varied from 0 to 1400 nematodes in 30 g/root. A survey was conducted in rural areas producing banana had shown that lesion nematodes were present in all areas where *P. coffeae* constituted only 3.2% of the nematode pest complex present in banana roots and 7.6% in soil samples [116].

4.5 Damage

The symptoms of damage caused by *P. coffeae* are very similar to those caused by *R. similis* as both are migratory endoparasites. They cause stunting of the plants, slow growth in vegetative phase, reduced number of leaves, lower bunch weight and reduce life span of plantations. There was reports of *P. coffeae* infections on banana plantations [107] which rendered the whole plantations unproductive.

4.5.1 Host range

The grapevine, citrus and veld (dune thicket, grasses) are the well-known hosts of *Pratylenchus coffeae* although it has a wide host range including many broad leaf weed species [101, 102] and information from both NCN and SAPPNS databases).

4.6 Management strategies

4.6.1 Legislation

In South Africa, when *R. similis* was discovered in banana under severe infection, the Government passed a legislation to prevent the spread of this nematode in new areas. This legislation is named as, "The Agricultural Pest Act 36 of 1983 [122]". According to this legislation, for the transport of planting material from one area to another area, a permit is required during transportation. In this act, the propagation materials are identified as suckers, rhizomes and setts. The tissue culture plants are the best technique to get rid of this nematode, because these plants are healthy and do not carry any plant pathogens but eventhough, a permit is required when any nursery is established there.

4.6.2 Preparation of plant material

In most developed countries, nematode free banana planting material is exclusively produced from tissue culture-based methods for commercial purposes.

However, in underdeveloped countries particularly in rural areas, suckers and rhizomes are still being used to produce propagative material especially where tissue culture-based methods are lacking. This results in spread of plant parasitic nematodes to healthy soil from infected sites. In such cases, paring is recommended where visible lesions caused by nematode pests and banana weevil is removed so that only white rhizome tissue remains. In many African countries especially in South Africa paring is generally followed by hot water treatment for a specific period of time to kill remaining nematode population and found to be effective in managing nematode pests. Once the rhizomes are treated, they must be planted in nematode free soil and for that the soil should be tested in nematological laboratory. This could be obtained by using one or combinations of the following strategies, like, keeping the field fallow for a certain period, organic amendments, soil sterilisation through heat by using transparent plastic cover for several weeks or planting the suckers in virgin soil. The soil sterilisation is not that much feasible in a large area as the use of polythene to a large area is not practical because of high input cost.

4.6.3 Cultural control

Cultural methods are the cheap and easily followed method and also eco friendly. These methods include, fallowing for at least six months, is very useful [123, 124] as this is generally used as monocropping and because of this practice, nematode population build up and cannot be controlled through any method. Another method under this is, crop rotation with selected cover crops before banana cultivation that helps in reducing the nematode population so that the new suckers will not get the infection from very beginning like, Milne and Keetch [121] tested several cover crops and reported that radish (*Raphanus sativus*) and *Tagetes patula* reduced populations of *R. similis* after 5 months compared to that of ethylene dibromide (EDB) fumigation. Rotation with Buffalo grass (*Megathyrsus maximus* var. *trichoglum*; syn *Panicum maximum*) and purple bean (*Phaseolus atropurpureus*) also reported to control *R. similis* and *Meloidogyne* spp. Sugarcane crop (*Saccharum* hybrid) eliminated *R. similis* after 10 weeks [118]. The third method is intercropping. Intercropping is the cultivation of some particular crops with the main crops to manage the nematodes. These crops may be coffee (*Coffea arabica*), vegetables, maize (*Zea mays*) and cassava (*Manihot esculenta*). The next method is incorporation of organic manures in large volume. These manures after decomposition, release some phenolic compounds which are harmful for the nematode survival [125]. Application of 15 tons chicken manure/ha or 30 tons cattle manure/ha is generally recommended before planting banana [118]. The organic amendments have multiple beneficial effects like, increase in plant growth by improving soil structure and fertility, improvement in plant resistance and the stimulation of micro-organisms, which act as natural enemies of nematodes [125]. In cases of severe nematode infections, treatment of banana plants with a nematicide is recommended.

4.7 Clean propagative materials

4.7.1 Banana tissue culture

To prevent the spread of pests and disease, use of tissue culture banana planting material is one of the best methods to avoid the nematode infection. The tissue cultured propagating material is grown in such a media that it is free from any disease. So, this is an important practice to get rid of any pathogen or nematode. Using these materials, the spread of nematodes and other pathogens is controlled from diseased field to healthy field. In Hawaii, most of the banana fields are infested with plant-parasitic nematodes [126], micro-propagation from disease free materials using sterile techniques, offers a good way to obtain nematode free planting materials.

4.7.2 Hot-water treatment

A hot water dip has been successfully used to control burrowing nematodes and root knot nematodes in anthurium and ginger, respectively. Although, various temperature-time combinations ranging from 5 min @ 50°C to 25 min @ 55°C are recommended by researchers across the world, CTAHR researchers recommend soaking of banana suckers at 50°C for 10 min for disinfestion.

4.7.3 Modified solarization

Soil solarization involves heating the soil using natural solar radiation beneath a transparent plastic sheet to reach lethal temperatures for soilborne pests. The method is effective against a range of soil inhabiting pests, pathogens and nematodes which live in the top 4 inches (10 cm) of soil. The nematodes in the deep layers escape from the lethal temperature attained by this method.

4.7.4 Biological control

Many biocontrol agents like, fungal and bacterial bioagents are beneficial for the management of nematode pests of banana. In 1998, Daneel et al. [127] has demonstrated the efficacy of the soil fungus, *Purpureocillium lilacinum* (syn *Paecilomyces lilacinus*) for the control of banana nematode pests including *R. similis* and *Meloidogyne* spp. This bioproduct was also responsible for reducing the period of growth from flowering to harvesting which is helpful in escaping the nematode problems. This product was used at a dosage rate of 2 × 109 spores/g in suspension at 2–4 g/mat, depending on the severity of nematode infestation [128]. It was registered for use in South Africa on banana. *P. lilacinus* is a common soil fungus used as biocontrol agents that has been isolated from many different habitats around the world. It acts as a facultative egg pathogen of sedentary nematodes and also an important option to control juvenile and adult burrowing nematodes in banana. Mendoza et al. [129] reported that this nematode antagonistic fungus may be used as an integrated approach to control plant parasitic nematodes of banana.

4.7.5 Corm destruction

Since the most damaging stages of nematodes spend a considerable amount of their life cycle inside roots, killing of banana plants using a non-selective herbicide i.e., Glyphosate simultaneously kills the obligate endoparasitic stages of the nematodes Thus greatly improving the potential of successive fallow to lower the nematode populations without using nematicides [130].

4.7.6 Cover-cropping

Since banana is a perennial crop, and farmers take the benefit of ratoon crop also within the same planting cost, it is difficult to manage nematodes over a longer period of time because the endoparasites are nearly impossible to manage even through chemicals. In this condition, growing of cover crop may become an option for the management. These cover crops release some chemicals which are having allelopathic compounds that are deleterious for the nematodes. Allelopathy is a biological phenomenon by which an organism produces one or more biochemicals that negatively affect the growth, survival, and reproduction of other organisms. The plants like, marigold (*Tagetes* spp.), sunhemp (*Crotalaria juncea*), rapeseed (*Brassica napus*, [131]), velvet bean (*Mucuna pruriens*, [132]), sorghum-sudan grass (*Sorghum bicolor* × *Sorghum arundinaceum* var. *Sudanense,* [133]), are reported as cover crops having allelopathy against plant parasitic nematodes. Among all these crops, marigold was found the best in banana cropping system and the allelopathy differs with different species of nematodes and marigold and also the soil temperature has the influence over it [134].

4.8 Production of healthy seedlings of banana

For the preparation of healthy banana suckers, a mixture of soil with biocontrol formulations and organic cakes can be prepared and used for hardening the seedlings. This mixture may include two kg each of *P. fluorescens* 1% W. P., *Trichoderma harzianum* 1% W. P. and *Purpureocillium lilacinus* 1% W. P. + five kg of carbofuran or phorate or 25 kg of neem cake or pongamia cake for preparing one ton of final mixture.

4.9 Chemical control

Conventionally, synthetic derived nematicides have been widely used for nematode control on banana. Although fumigants have been highly effective [135], such products are not used in banana production mainly due to high input costs and now these are banned also to be used in agriculture. The carbamates and the organophosphates are used regularly in the banana cultivation as pre and post application. These chemicals are used as granular or liquid formulations. During application, these are sprayed around the base of pseudostems or suckers. In South Africa, a chemical, Furfuraldehyde is registered for banana and it can be used when the population of nematodes is below economic damage level [128]. Because of unawareness and hidden mode of life cycle of plant parasitic nematodes, they are not given so much importance although they cause sufficient loss in the yield. Therefore, it is recommended that nematode samples are taken annually for nematode population estimation and that nematicides are only applied to reduce nematode pest populations likely to limit yield or cause long term yield decline. Although pre plant treatments such as soil fumigation with Telone II® (1,3-dicloropropene) are very effective in suppressing nematode populations, such treatments are short lived compared to the life of a banana plot. The following treatments should be done to manage the nematodes in banana:

1. Use of tissue cultured (*in vitro* produced) plants.

2. Rotation with alternative crops for minimum of 2 years.

3. Fallow in the absence of banana ‘volunteers’ for 10–12 months.

4. Selection of disease-free suckers.

5. Paring of diseased tissue from corms.

6. Immersing suckers in hot water for a particular time and period.

7. Flooding for 8 weeks after having destroyed previous banana crop.

8. Applying a nematicide to planting hole and in fill soil.

9. Regular spot applications with nematicides.

4.10 Practices that maintain productivity and vigour

1. Support plants with bamboo poles or with string or ropes to prevent plants toppling.

2. Regular application of mulches of grass, leaves or organic waste.

3. Grow cultivars with robust stature and wind tolerance and endophytes [136] including *Trichoderma viride* and non-pathogenic *F. oxysporum* [137, 138].

5. Conclusions

In banana cultivation, plant parasitic nematodes are not causing much damage or the loss and is not that much economical but there is yield reduction and the production cost increases very high when compared to the yield. Nowadays in banana cultivation, tissue culture propagation became a common practice, so, after planting in an orchard, there is rare chance of getting infection of plant parasitic nematodes in the first year. The cultivators also become very aware about the monitoring and testing the soils before planting and after planting every year. After getting tested from the designated authorities, they follow the recommendations given by them. Thus, these producers manage effectively the infestation of nematode problems. But, we have to be very careful and always be ready with some alternative management practices that could be followed if found any severe infection. The need for the management is more necessary in small holding farmers and growers because of their small holding size.

Author details

Nishi Keshari* and Gurram Mallikarjun
Department of Nematology, Dr. Rajendra Prasad Central Agricultural University, Pusa, Samastipur, Bihar, India

*Address all correspondence to: nishinema@gmail.com

References

[1] Anonymous. Citrus Fruit Fresh and Processed Annual Statistics 2002. Rome, Italy: Food and Agriculture Organization of the United Nations; 2002

[2] Ford HW. The influence of rootstock and tree age on root distribution of citrus. Proceedings of the American Society for Horticultural Science. 1954;**63**:137-172

[3] Ford HW. Root distribution in relation to the water table. Proceedings of the Florida State Horticultural Society. 1954;**67**:30-33

[4] Duncan LW. Nematode parasites of citrus. In: Luc M, Sikora RA, Bridge J, editors. Plant Parasitic Nematodes in Subtropical and Tropical Agriculture. Wallingford, UK: CAB International; 2005. pp. 437-466

[5] Heald CM, O'Bannon JH. Citrus Declines Caused by Nematodes. V. Slow Decline. Florida Department of Agriculture and Consumer Services, Division of Plant Industry, Nematology Circular No. 143; 1987

[6] Van Gundy SD, Meagher JW. Citrus nematode (*Tylenchulus semipenetrans*) problems worldwide. In: 1977 International Citrus Congress; Orlando, Florida. 1977

[7] Cobb NA. The citrus root knot nematode. Journal of Agricultural Research. 1914;**2**:217-230

[8] Siddiqi MR. Occurrence of the citrus nematode, *Tylenchulus semipenetrans* Cobb, 1913, and the reniform nematode, *Rotylenchulus reniformis* in India (Abstr.). In: Proceedings of 48th Indian Science Congress Part III. 1961. p. 504

[9] Timmer LW, Garnsey SM, Broadbent P. Diseases of citrus. In: Diseases of Tropical Fruit Crops. Wallingford, UK: CAB International; 2003. pp. 163-196

[10] Shokoohi E, Duncan LW. Nematode parasites of citrus. In: Sikora R, Timper P, Coyne D, editors. Plant-Parasitic Nematodes in tropical & Subtropical Agriculture. 3rd ed. St. Albans, UK: CAB International; 2018. pp. 446-476

[11] Abd-Elgawad MMM, Abou-Deif MH, Hammam MMA, Abd-El-Khair H, Koura FFH, Abd El-Wahab AE, et al. Effect of infection with *Tylenchulus semipenetrans* on enzymatic activities in citrus. International Journal of Engineering and Innovative Technology. 2015;**4**(12):43-48

[12] Van Gundy SD. The life history of the citrus nematode, *Tylenchulus semipenetrans* Cobb. Nematologica. 1958;**3**(4):283-294

[13] Duncan LW. Managing nematodes in citrus orchards. In: Ciancio A, Mukerji KG, editors. Integrated Management of Fruit Crops and Forest Nematodes. Springer Netherlands: Springer Science+Business Media B.V; 2009. pp. 135-173. DOI: 10.1007/978-1-4020-9858-1

[14] Cohn E. The development of the citrus nematode on some of its hosts. Nematologica. 1965;**11**:593-600

[15] Vilardebo A, Luc M. Slow decline of citrus caused by the nematode, *Tylenchulus semipenetrans*. Fruits. 1961;**16**:261-261

[16] Baines RC, Miyakawa T, Cameron JW, Small RH. Infectivity of two biotypes of the citrus nematode on citrus and on some other hosts. Journal of Nematology. 1969;**1**:150-159

[17] Inserra RN, Duncan LW, O'Bannon JH, Fuller SA. Citrus Nematode Biotypes and Resistant Citrus Rootstocks in Florida. Nematology Circular No. 205. Florida Department of Agriculture and Consumer Services, Division of Plant Industry; 1994

[18] Stokes DE. *Andropogon rhizomatus* parasitized by a strain of *Tylenchulus semipenetrans* not parasitic to four citrus rootstocks. Plant Disease Reporter. 1969;**53**:882-885

[19] Thorne G. Principles of Nematology. New York, NY: McGraw-Hill Book Company, Inc.; 1961

[20] Parvatha Reddy P, Singh DB. Evaluating the reaction of some species and varieties of *Citrus* and *Poncirus* to the citrus nematode. Indian Journal of Nematology. 1978;**8**:82-84

[21] El-Mohamedy RSR, Hammam MMA, Abd-El-Kareem F, Abd-Elgawad MMM. Biological soil treatment to control *Fusarium solani* and *Tylenchulus semipenetrans* on sour orange seedlings under greenhouse conditions. International Journal of ChemTech Research. 2016;**9**(7):73-85

[22] Abobatta WF. Citrus varieties in Egypt: An impression. International Research Journal of Applied Sciences. 2019;**1**:63-66

[23] Hammam MMA, El-Mohamedy RSR, Abd-El-Kareem F, Abd-Elgawad MMM. Evaluation of soil amended with bio-agents and compost alone or in combination for controlling citrus nematode *Tylenchulus semipenetrans* and fusarium dry root rot on Volkamer lime under greenhouse conditions. International Journal of ChemTech Research. 2016;**9**(7):86-96

[24] Safdar A, Javed N, Khan SA, Safdar H, Haq IU, Abbas H, et al. Synergistic effect of a fungus, *Fusarium semitectum*, and a nematode, *Tylenchulus semipenetrans*, on citrus decline. Pakistan Journal of Zoology. 2013;**45**(3):643-651

[25] Nemec S, Phelps D, Baker R. Effects of dihydrofusarubin and isomarticin from *Fusarium solani* on carbohydrate status and metabolism of rough lemon seedlings. Phytopathology. 1989;**79**(6):700-705

[26] Labuschange N, van der Vegte FA, Koteze J.M. Interaction between *Fusarium solani* and *Tylenchulus semipenetrans* on citrus roots. Phytophylactica. 1989;**21**(1):29-33

[27] Abd-Elgawad MMM, Koura FFH, Montasser SA, Hammam MMA. Distribution and losses of *Tylenchulus semipenetrans* in citrus orchards on reclaimed land in Egypt. Nematology. 2016;**18**:1141-1150. Available from: http://journals.indexcopernicus.com/EgyptianJournalof Agronematology, p8230,3.html

[28] Abd-Elgawad MMM, McSorley R. Movement of citrus nematode-infested material onto virgin land: Detection, current status and solutions with cost-benefit analysis for Egypt. Egypt Journal of Agronematology. 2009;**7**(1): 35-48

[29] Duncan LW, Graham JH, Timmer LW. Seasonal patterns associated with *Tylenchulus semipenetrans* and *Phytophthora parasitica* in the citrus rhizosphere. Phytopathology. 1993;**83**:573-581

[30] Galeano M. Dinamica di popolazione di *Tylenchulus semipenetrans* e della nematofauna di vita libera nella rizosfera di agrumi in Spagna. Supplemento Nematologia Mediterranea. 2002;**30**: 49-53

[31] Sorribas FJ, Verdejo-Lucas S, Forner JB, Alcaide A, Pons J, Ornat C. Seasonality of *Tylenchulus semipenetrans* Cobb and *Pasteuria* sp. in citrus orchards in Spain. Supplement to the Journal of Nematology. 2000;**32**(4S):622-632

[32] Siddiqui MA, Alam MM. Toxicity of different plant parts of *Tagetes lucida* to plant parasitic nematodes. Indian Journal of Nematology. 1988;**18**:181-185

[33] Park JO, Hargreaves JRE, McConville J, Stirling GR, Ghisal-berti EL, Sivasithamparam K. Production of

leucinostatins and nematicidal activity of Australian isolates of *Paecilomyces lilacinus* (Thom) Samson. Letters in Applied Microbiology. 2004;**38**:271-276

[34] Mclean KL, Dodd SL, Sleight BE, Hill RA, Stewart A. Comparison of the behavior of a transformed hygromycin resistant strain of *Trichoderma viride* with the wild type strain. New Zealand Plant Protection. 2004;**57**:72-76

[35] Ali Ayat M. Control of strawberry fungal diseases under organic agriculture system [Ph.D. thesis]. Egypt: Fac. Agric. Cairo Univ.; 2013. 124 p

[36] Huang XW, Zhao NH, Zhang KQ. Extracellular enzymes serving as virulence factors in nematophagous fungi involved in infection of the host. Research Microbiology. 2005;**115**:811-816

[37] Huang Y, Xu C, Ma L, Zhang K, Duan C, Mo M. Characterization of volatiles produced from Bacillus megateriumYFM3.25 and their nematicidal activity against *Meloidogyne incognita*. European Journal of Plant Pathology. 2009;**26**:417-422

[38] Khan MR, Kounsar K, Hamid A. Effect of certain rhizobacteria and antagonistic fungi on root-knot nodulation and root knot nematode disease of green gram. Nematologia Mediterranea. 2002;**30**(1):85-89

[39] Mahadevan B, Crawford DL. Properties of the chitinase of the antifungal biocontrol agent Streptomyces lydicus WYEC108. Enzyme and Microbial Technology. 1997;**20**:489-493

[40] Nemec S, Datnoff LE, Strandberg J. Efficacy of biocontrol agents in planting mixes to colonize plant roots and control root diseases of vegetables and citrus. Crop Protection. 1996;**15**:735-742

[41] Dicklow MB, Acosta N, Zuckerman BM. A novel *Streptomyces* species for controlling plant parasitic nematodes. Journal of Chemical Ecology. 1993;**19**:159-173. DOI: 10.1007/BF00993686

[42] Le Roux HF, Pretorius MC, Huisman L. Citrusnematode IPM in Southern Africa. Proceedings of the International Society of Citriculture. 2000;**2**:823-827

[43] Coley-Smith JR. Some interaction in soil between plants, sclerotium forming fungi and other microorganisms. In: Friend J, Threlfall DR, editors. Biochemical aspects of Plant Parasite Relationships. London, New York, San Fracisco: Academic Press; 1976. pp. 11-23

[44] Giannakou IO, Karpouzas DG, Anastasiades I, Tsiropoulos NG, Georgiadou A. Factors affecting the efficacy of non-fumigant nematicides for controlling root-knot nematodes. Pest Management Science. 2005;**61**:961-972

[45] Le Roux HF, Ware AB, Pretorius MC. Comparative efficacy of preplant fumigation and postplant chemical treatment of replant citrus trees in orchards infested with *Tylenchulus semipenetrans*. Plant Disease. 1998;**82**:1323-1327

[46] O'Bannon JH, Tarjan AC. Preplant fumigation forcitrus nematode control in Florida. Journal of Nematology. 1973;**5**:88-95

[47] Reynolds HW, O'Bannon JH. Factors influencing the citrus nematode and its control on citrus replants in Arizona. Nematologica. 1963;**9**:337-340

[48] Sorribas FJ, Verdejo-Lucas S, Galeano M, Pastor J, Or-nat C. Effect of 1,3-dichloropropene and rootstocks alone and incombination on *Tylenchulus semipenetrans* and citrus tree growth in areplant management program. Nematropica. 2003;**34**:149-158

[49] Davis RM, Heald CM, Timmer LW. Chemical control of the citrus nematode

on grapefruit. Journal of the Rio GrandeValley Horticultural Society. 1982;**35**:59-61

[50] Van Gundy S, Garabedian S, Nigh EL. Alternatives to DBCP for citrus nematode control. Proceedings of the International Society of Citriculture. 1982;**1**:387-390

[51] Wheaton TA, Childers CC, Timmer LW, Duncan LW, Nikdel S. Effects of aldicarb on yield, fruit quality, and tree condition on Florida citrus. Proceedings of the Florida State Horticultural Society. 1985;**98**:6-10

[52] Le Roux HF, Ware AB. Accelerated degradation ofsome soil-applied nematicides in a South African citrus orchard. Pro-ceedings International Society of Citriculture. 1996;**1**:593-596

[53] McClure MA, Schmitt ME. Control of citrus nema-tode, *Tylenchulus semipenetrans*, with cadusafos. Supplement to Journal of Nematology. 1996;**28**:624-628

[54] Philis J. Effect of citrus nematode control on the yield and fruit quality of grapefruit in Cyprus. Miscellaneous Report. 1997;**66**:3-6

[55] Walker GE, Morey BG. Effects of chemicals andmicrobial antagonists on nematodes and fungal pathogens of citrus roots. Australian Journal of Experimental Agriculture. 1999;**39**: 629-637

[56] Galeano M, Verdejo-Lucans S, Sorribas FJ, Ornat C, Forner JB, Alcaide A. New citrus selections from Cleopatra mandarin x *Poncirus trifoliate* with resistance to *Tylenchulus semipenetrans* Cobb. Nematology. 2003;**5**:227-234

[57] Kaplan DT. Characterization of citrus rootstock responses to *Tylenchulus semipenetrans* (Cobb). Journal of Nematology. 1981;**13**:492-498

[58] Ling P, Duncan LW, Deng Z, Dunn D, Hu X, Huang S, et al. Inheritance of citrus nematode resistance and its linkage with molecular markers. Theoretical Applied Genetics. 2000;**100**:1010-1017

[59] Lee RF, Lehman PS, Navarro L. Nursery practices and certification programs for budwood and rootstocks. In: Citrus Health Management. St. Paul, MN: APS Press; 1999. pp. 35-46

[60] Edwards M. Resistance and tolerance of grapevine rootstocks to plant-parasitic nematodes in vineyards in North-East Victoria. Australian Journal of Experimental Agriculture. 1989;**29**:129-131

[61] Duncan LW, Inserra RN, O'Bannon JH, El-Morshedy MM. Reproduction of a Florida population of *Tylenchulus semipenetrans* on resistant citrus rootstocks. Plant Disease. 1994;**78**:1067-1071

[62] Lehman PS. Role of plant protection organizations in nematode management. XIX Congress of Brazilian Society of Nematology; Rio Quente, Brazil; 1996. pp. 137-148

[63] Cronjé C, Le Roux HF, Truter M, Van Heerden I, Phillips H. Long-term effect of preplant soil solarisation on growth of replant citrus trees in South Africa. African Plant Protection. 2002;**8**:41-49

[64] Stapleton JJ, Elmore CL, DeVay JE. Solarization and biofumigation help disinfest soil. California Agriculture. 2000;**54**:42-45

[65] Tjamos EC, Grinstein A, Gamliel A. Disinfestation of soil and growth media. In: Albajes R, Gullino ML, van Lenteren JC, Elad Y, editors. Integrated Pest and Disease Management in Greenhouse Crops. Dordrecht, The Netherlands: Kluwer Academic; 1999. pp. 139-149

[66] Baines RC, Klotz LJ, Clarke OF, DeWolfe TA. Hot water treatment of orange trees for eradication of citrus nematode. California Citrograph. 1949;**34**:482-484

[67] Silva HP, Montero AR, Feraz LCCB. Trata-mento hidrotérmico de mudas de cı́tricos para a erradicaçãode *Tylenchulus semipenetrans*. Nematologia Brasileira. 1987;**11**:143-152

[68] Ibrahim IKA, Mokbel AA, Handoo ZA. Current status of phytoparasitic nematodes and their host plants in Egypt. Nematropica. 2010;**40**: 239-262

[69] Inserra RN, Duncan LW, Troccoli A, Dunn D, Maia SosSantos J, Vovlas N. *Pratylenchus jaehni* sp. n. from citrus in Brazil and a redescription of *P. coffeae*. Nematology. 2001;**3**:653-665

[70] O'Bannon JH, Tomerlin AT. Population studies on two species of *Pratylenchus* on citrus. Journal of Nematology. 1969;**1**:299-300

[71] O'Bannon JH, Tomerlin AT. Citrus tree decline caused by *Pratylenchus coffeae*. Journal of Nematology. 1973;**5**:311-316

[72] De Campos AS, dos Santos JM, Duncan LW. Nematodes of citrus in open nurseries and orchards in Sao Paulo State, Brazil. Nematology. 2002;**4**:263-264

[73] Tomerlin AT, O'Bannon JH. Effect of *Radopholus similis* and *Pratylenchus brachyurus* on citrus seedlings in three soils. Soil and Crop Science Society of Florida Proceedings. 1974;**33**:95-97

[74] Inserra RN, Vovlas N. Effects of *Pratylenchus vulnus* on the growth of sour orange. Journal of Nematology. 1977;**9**:154-157

[75] Siddiqi MR. Studies on nematode root rot of citrus in Uttar Pradesh, India. Proceedings of Zoological society, Calcutta. 1964;**17**:67-75

[76] Radewald JD, O'Bannon JH, Tomerlin AT. Anatomical studies of *Citrus jambhiri* roots infected by *Pratylenchus coffeae*. Journal of Nematology. 1971;**3**: 409-416

[77] Baghel PPS, Bhatti DS. Evaluation of pesticides for the control of phytonematodes on citrus. In: Third Nematology Symposium. Solan: Himachal Pradesh Agricultural University; 1983. pp. 38-39

[78] Van Gundy SD, Thomason IJ, Rackham RL. The reaction of three *Citrus* spp. to three *Meloidogyne* spp. Plant Disease Reporter. 1959;**43**:970-971

[79] Whitehead AG. Taxonomy of *Meloidogyne* (Nematoda: Heteroderidae) with descriptions of four new species. Transactions of the Zoological Society of London. 1968;**31**:263-401

[80] Patel D, Patel BA, Patel SK, Patel RL, Patel RG. Root knot nematode, *Meloidogyne indica* on kagzi lime in North Gujarat. Indian Journal of Nematology. 1999;**29**:185-205

[81] Lazan H, Ali ZM. Guava. In: Shaw PE, Chan HT Jr, Nagy S, editors. Tropical and Subtropical Fruits. Auburndale, Florida: AgScience, Inc.; 1998. pp. 446-485

[82] Khan RM, Kumar S, Reddy PP. Role of plant parasitic nematode(s) and fungi in guava wilt. Pest Management in Horticultural Ecosystems. 2001;**7**:152-161

[83] Mahto Y, Edward JC. Studies on pathogenicity, host parasite relationship and histopathological changes of some important fruit trees due to predominant phytonematode associated with them (Part 1). Allahabad Farmer. 1979;**50**:403

[84] Nigam K, Verma RS, Verma AK, Sinha V. Pathogenicity of Hoplolaimus spp. Daday, 1905 to guava (*Psidium gujava*). Advances in Agricultural Research in India. 1995;**3**:158-160

[85] Abivardi C. A stylet nematode, *Tylenchorhynchus cylindricus* Cobb 1913, infesting the common guava, *Psidium guajava* L. in Iran. Nematologia Mediterranea. 1973;**1**:139-140

[86] Mokbel AA. Nematodes and their associated host plants cultivated in Jazanprovince, Southwest Saudi Arabia. Egyptian Journal of Experimental Biology (Zoology). 2014;**10**:35-39

[87] Gomes VM, Souza RM, Silva MM, Dolinski C. Nutritional status of guava (*Psidium guajava* L.) plants parasitized by *Meloidogyne mayaguensis*. Nematologia Brasileira. 2008;**32**:154-160

[88] Cetintas R, Kaur R, Brito JA, Mendes ML, Nyczepir AP, Dickson DW. Pathogenicity and reproductive potential of *Meloidogyne mayaguensis* and *M. floridensis* compared with three common *Meloidogyne* species. Nematropica. 2007;**37**:21-31

[89] Rao MS. Papaya seedlings colonized by the bio-agents, *Trichoderma harzianum* and *Pseudomonas fluorescens* to control root-knot nematodes. Nematologia Mediterranea. 2007;**35**:199-203

[90] Jindapunnapat K, Chinnasri B, Kwankuae S. Biological control of root knot nematodes (*Meloidogyne enterolobii*) in guava by the fungus, *Trichoderma harzianum*. Journal of Developments in Sustainable Agriculture. 2013;**8**:110-118

[91] Al-Hazmi AS, Dawabah AAM, Al-Nadhari SN. *Verticillium chlamydosporium*, a fungal parasite of the cereal cyst nematode (*Heterodera avenae*) in the Saudifields. In: The 4th International Cereal Nematodes Initiative Workshop. 22-24 Aug., 2013. Beijing: Friendship Hotel; 2013

[92] Rao MS. Effect of combinations of bio-pesticides on the management of nematodes on *Carica papaya* L. Acta Horticulturae. 2010;**1**:459-464

[93] Oka Y. Mechanisms of nematode suppression by organic soil-amendments—A review. Applied Soil Ecology. 2010;**44**:101-115

[94] Rodriguez-Kabana R. Organic and inorganic nitrogen amendments to soilas nematode suppressants. Journal of Nematology. 1986;**18**:129-135

[95] Gomes VM, Souza RM, Corrêa FM, Dolinski C. Management of *Meloidogyne mayaguensis* in commercial guava orchards with chemical fertilization and organic amendments. Nematologia Brasileira. 2010;**34**:23-30

[96] Alam MM. In: Nemtol PJ, editor. Effect of Ammonia on the Population of Plant Parasitic Nematodes and Growth of Some Vegetables. Vol. 10. 1992. pp. 133-137

[97] Al-Hazmi AS, Dawabah AAM. Effect of urea and certain NPK fertilizers onthe cereal cyst nematode (*Heterodera avenae*) on wheat. Saudi Journal of Biological Sciences. 2014;**21**:191-196

[98] Al-Hazmi AS, Dawabah AAM, Al-Nadhari SN, Al-Yahya FA. Comparativeefficacy of different approaches to managing *Meloidogyne incognita* on green bean. Saudi Journal of Biological Sciences. 2017;**24**: 149-154

[99] Akhtar M, Malik A. Roles of organic soil amendments and soil organisms inthe biological control of plant-parasitic nematodes: A review. Bioresource Technology. 2000;**74**:35-47

[100] McSorley R. Overview of organic amendments for management of plant-parasitic nematodes, with case studies from Florida. Journal of Nematology. 2011;**43**:69-81

[101] Gowen S, Quénéhervé P. Nematode parasites of bananas and plantains. In: Luc M, Sikora RA, Bridge J, editors. Plant Parasitic Nematodes in

Subtropical and Tropical Agriculture. Wallingford: AB International; 2005. pp. 611-643

[102] Kleynhans KPN, Van den Berg E, Swart A. Plant nematodes in South Africa. Plant Protection Research Institute Handbook No 8. Pretoria: Agricultural Research Council–Plant Protection Research Institute; 1996

[103] Jones RK, Milne DL. Nematode pests of bananas. In: Keetch DP, Heyns J, editors. Nematology in Southern Africa. Science Bulletin No. 400. Pretoria: Department of Agriculture and Fisheries; 1982. pp. 30-37

[104] De Jager K, Daneel MS, Desmet M, et al. Pathogenicity and distribution studies to determine threshold levels for nematodes on banana. Banana Growers Association of South Africa. 1999;**3**:72-77

[105] Gowen SR, Queneherve P. Nematode parasites of bananas, plantains and abaca. In: Plant Parasitic Nematodes of Subtropical and Tropical Agriculture. Walligford, UK: CAB International; 1990. pp. 431-460

[106] Daneel MS, De Jager K, Van den Bergh I, et al. Occurrence and pathogenicity of plant-parasitic nematodes on commonly grown banana cultivars in South Africa. Nematropica. 2015;**45**:118-127

[107] Willers P, Daneel MS, De Jager K. Banana. In: Van den Berg MA, De Villiers EA, Joubert PH, editors. Pest and Beneficial Arthropods of Tropical and Non-Citrus Subtropical Crops in South Africa. Nelspruit: Agricultural Research Council–Institute for Tropical and Subtropical Crops; 2001. pp. 34-43

[108] Willers P. Nematologiese navorsing in subtropiese bedrywe. Neltropica Bull. 1998;**299**:12-13

[109] Sher SA. Revision of the Genus *Radopholus* Thome, 1949 (Nematoda, Tylenchoidea). Proceedings of the Helminthological Society of Washington. 1968;**35**:219-237

[110] Marin DH, Sutton TB, Barker KR. Dissemination of bananas in Latin America and the Caribbean and its relationship to the occurrence of *Radopholus similis*. Plant Disease. 1998;**82**:964-974

[111] Milne DL, Kuhne FA. Nematodes attack bananas. Farming in South Africa. 1968;**44**:5-9

[112] Rabie EC. Die invloed van *Meloidogyne javanica* en *M. incognita* op die voorkoms van Valspanamasiekte by piesangs [MSc dissertation]. Pretoria: University of Pretoria; 1991

[113] Jonathan EI, Gajendran G, Manuel WW. Management of *Meloidogyne incognita* and *Helicotylenchus multicinctus* in banana with organic amendments. Nematologia Mediterranea. 2000;**28**:103-105

[114] Kim J, Seo SM, Lee SG, Shin SC, Park IK. Nematicidal activity of plant essential oils and components from coriander *(Coriandrum sativum)*, oriental sweetgum (*Liquidambar orientalis*), and valerian (*Valeriana wallichii*) essential oils against pinewood nematode (*Bursaphelenchus xylophilus*). Journal of Agricultural and Food Chemistry. 2008;**56**(16):7316-7320

[115] Seenivasan N. Bio-efficacy of anti-nemic plants against root knot nematode in medicinal coleus. Journal of Eco-Friendly Agriculture. 2011;**6**(1):92-96

[116] Daneel MS, Dillen N, Husselman J, et al. Results of a survey on nematodes of *Musa* in household gardens in South Africa and Swaziland. InfoMusa. 2003;**12**:8-11

[117] Quénéhervé P, Cadet P. Localisation des nematodes dans les rhizomes du

bananier cv Poyo. Revue de Nématologie. 1985;**8**:3-8

[118] De Villiers EA, Daneel MS, Schoeman PS. Pests. In: Robinson JC, De Villiers EA, editors. The Cultivation of Banana. Nelspruit: Ingwe Print; 2007. pp. 194-219

[119] Duchame EP, Birchfield W. Physiologic races of the burrowing nematode. Phytopathology. 1956;**46**: 615-616

[120] Keetch DP. Some host plants of the burrowing eelworm, *Radopholus similis* (Cobb) in Natal. Phytophylactica. 1972;**4**:51-58

[121] Milne DL, Keetch DP. Some observations on the host plant relationships of *Radopholus similis* in Natal. Nematropica. 1976;**6**:13-17

[122] NDA. 2015. Available from: http://www.nda.agric.za/docs/NPPOZA/Agricultural%20Pests%20Act.pdf

[123] Loos CA. Eradication of the burrowing nematode, *Radopholus similis*, from bananas. Plant Disease Report. 1961;**45**:457-461

[124] Tarjan AC. Longevity of *Radopholus similis* (Cobb) in host free soil. Nematologica. 1961;**6**:170-175

[125] Stirling GR. Biological Control of Plant Parasitic Nematodes. Progress, Problems and Prospects. Wallingford: CAB International; 1991

[126] Wang K-H, Hooks CRR. Survey of nematodes on banana in Hawaii and methods used for their control. In: CTAHR Cooperative Extension Service PD-69. 7 pp. 2009. Available from: http://www.ctahr.hawaii.edu/oc/freepubs/pdf/PD-69.pdf

[127] Daneel MS, De Jager K, Dreyer S. PL Plus is an environmentally friendly nematicide for banana nematodes. Neltropica Bulletin. 1998;**300**:32-34

[128] Van Zyl K. A guide to crop pest management in South Africa. A compendium of acaricides, insecticides, nematicides, molluscicides, avicides and rodenticides. In: A Crop Life Compendium. 1st ed. Pinetown: VR Print; 2013

[129] Mendoza A, Sikora RA, Kiewnick S. Efficacy of *Paecilomyces lilacinus* strain 251 for the control of *Radopholus similis* in banana. Communications in Agricultural and Applied Biological Sciences. 2004;**69**:365-372

[130] Chabrier C, Quénéhervé P. Control of the burrowing nematode (*Radopholus similis* Cobb) on banana, impact of the banana field destruction method on the efficiency of the fallowing fallow. Crop Protection. 2003;**22**:121-127

[131] Wang KH, Sipes BS, Schmitt DP. Suppression of *Rotylenchulus reniformis* by *Crotalaria juncea*, *Brassica napus*, and *Target erecta*. Nematropica. 2001;**31**: 237-251

[132] Zasada IA, Klassen W, Meyer SLF, Codallo M, Abdul-Baki AA. Velvetbean (*Mucuna pruriens*) extracts: Impact on *Meloidogyne incognita* survival and on *Lycopersicon esculentum* and *Lactuca sativa* germination and growth. Pest Management Science. 2006;**62**:1122-1127

[133] Widmer TL, Abawi GS. Mechanism of suppression of *Meloidogyne hapla* and its damage by a green manure of Sudan grass. Plant Disease. 2000;**84**:562-568

[134] Ploeg AT, Maris PC. Effect of temperature on suppression of *Meloidogyne incognita* by *Tagetes* cultivars. Journal of Nematology. 1999;**31**:709-714

[135] Keetch DP, Reynolds RE, Mitchell JA. An evaluation of pre- and post–plant nematicides for the control of plant parasitic nematodes on bananas. Citrus and Subtropical Fruits Journal. 1976;**506**:5-7

[136] Sikora RA, Schuster RP. Novel approaches to nematode IPM. In: Frison EA, Gold CS, Karamura EB, Sikora RA, editors. Mobilizing IPM for Sustainable Banana Production in Africa. Montpellier, France: INIBAP; 1998. pp. 127-136

[137] Felde AZ, Pocasangre L, Sikora R. The potential use of microbial communities inside suppressive banana plants to increase root health and suppression of the burrowing nematode, *Radopholus similis*. In: Proceeding of the International Symposium—Banana Root Systems, Towards a Better Understanding for its Productive Management; 3-5 November 2003; Costa Rica. Corbana–INIBAP; 2004

[138] Niere B, Gold CS, Coyne D. Can fungal endophytes control soilborne pests in banana? In: Sikora RA, Gowen S, Hauschild R, Kiewnick S, editors. Multitrophic Interactions in Soil and Integrated Control, IOBC/WPRS Bulletin. Vol. 27. 2004. pp. 203-210

Chapter 5

Management of Root-Knot Nematode, *Meloidogyne Incognita* Dreaded Invading in Pointed Gourd (*Trichosanthes dioica* Roxb.) Crop Prone to Eastern U.P of India

Ali Anwar, Najeeb Mohammad Mughal, Efath Shahnaz, Saba Banday, Taibah Bashir, Qadrul Nisa and Gulam Jeelani

Abstract

Pointed gourd belongs to cucurbitaceae family and is extensively cultivated in eastern Uttar Pradesh (10000 Hectares), Bihar (14000 hectares), West Bengal, Assam, Orissa, Madhya Pradesh, Maharashtra and Gujrat. Its plants are perennial in nature and can survive for several years even if left uncared. This crop occupies large area of land in India. The system of cultivation varies from region to region such as trained on pandals or ardours especially during the rainy season in southern and western India. However, it is most susceptible to root-knot nematode, *Meloidogyne incognita*, the population level or density of root-knot nematodes were found in the range of 15–100 per cent of the root and soil samples. This nematode induces severe damage to pointed gourd on coarse-textured sandy soils, particularly during droughts stress. Crop failure is noticed at earlier stage of vines. In view of fact it is necessary to evolve the integrated strategies for management of root-knot nematode in this viny crop.

Keywords: Pointed Gourd, Root-Knot Nematode, Integrated management

1. Introduction

Pointed gourd (*T. dioica,Roxb.),* locally known as parwal in India is a staple vegetable of people of India and frequently used in various cuisines. Belonging to the Cucurbitaceae family, fruits of pointed gourd can be either oblong or rounded and mostly recognised with white to yellow stripings on the outer skin. The veggie has white and mushy flesh and used in various cuisines. Parwal/pointed gourd is not only consumed as a fresh vegetable but possess proven medicinal value. It is said to be useful in disorders of the circulatory system. Parwal leaves with the bark of *Azadirachta indica* are used for the treatment of leprosy. The nutritive value of parwal as reported by [1, 2] is as protein 2 g, fat 0.3 g; mineral 0.5 g, carbohydrate 2.2 g, calcium 30 mg, phosphorus 40 mg. iron 1.70 mg, carotene 153 mg, thiamine 0.05 mg, riboflavin 0.06 mg vitamin 2.0 g. Its crunchy seeds are also edible.

In India, it is stir-fried, used in stews, soups, and meat dishes. It is full fill maximum nutritional requirements of human vegetable diets which are rich in several important vitamins and minerals including it in a brilliant way to stay active and fit. Here are some of its popular health benefits.

- It is rich in fibre and promotes good digestive health by treating ailments in the digestive system.

- It improves the immunity of the body and prevents you from catching regular flu, cold and sore throat.

- According to Ayurveda, Pointed gourd is a natural blood purifier and filters out all the toxins and impurities.

- Its seeds help in controlling blood sugar levels and protect you from the verge of becoming a diabetes patient

Pointed gourd (*T. dioica*) is extensively cultivated in eastern India. Substantially, the habit of crop is perennial and vegetatively propagated through the cuttings and root-suckers, while seed propagation is avoided due to poor germination. The crop fetches more prices in the market and its demand due to everyday consumption as vegetable and stupendous nutritive value. A preliminary survey of the crop in eastern U.P. India indicated the association of root knot nematode with unthrifty growth of plants in many areas [3, 4]. It is extensively grown in eastern part of Uttar Pradesh (10000 ha), Bihar (14000 ha), West Bengal (25000 ha) and to some extent in Assam (5000 ha), Orissa, Madhya Pradesh, Maharashtra and Gujarat states in India. This crop is viny in nature and occupies large areas and hence the system of cultivation varies region wise such as trained on bamboo stack or arbours especially during rainy season. The plants of pointed gourd are perennial in habit and can survive for several years even if left uncared.

It is most susceptible to many pest and diseases resulting heavy loss of fruit yield reducing the income of marginal farmers whose are mostly cultivated in India and elsewhere. Among these, nematodes cause severe losses to pointed gourd. The extent in production of crop fruit yield by phytoparasitic nematodes depends to a large extent on the farming system employed. In general nematodes may be less injurious to the plants under more extensive and varied growing systems i.e. multiple crop farming and shifting or staking cultivation in subsistence agriculture

Figure 1.
Crop infected with root- knot nematode.

DOI: http://dx.doi.org/10.5772/intechopen.98923

or in widely spaced rotations of commercial farming systems than in more intensive production where more cropping and narrow rotations are practiced [5].

Perennial cropping systems, promote nematode population build up with time. The extent of the increase depends on the nematodes initially present and on the percentage of susceptible plant per unit area. Intensity of damage usually increases slowly with time in the perennial cropping system, as compared to the rapid increase in damage encounter in large scale parwal production where near annual cropping is practiced (**Figure 1**).

1.1 Noxious threat to pointed gourd

Root-knot nematode caused by *M.incognita* is a serious problem associated with field production of pointed gourd and cause major losses in crop in commercial farms, green houses and home gardens elsewhere. This disease is worldwide in distribution, essentially occurring in the area where hot summer is long but winter is short and mild. Nevertheless, it is not confined only to tropics and sub-tropics. It is also found in temperate regions.

In a fortified survey conducted by [6–8] of pointed gourd cultivated area of the farmer fields of eastern U.P. and another areas of the state showed that the population level of root-knot nematode (*Meloidogyne incognita*) was the range of 15–100 J2 per cent in root and soil samples.

Root-knot nematode (*M.incognita*) induces severe damage to pointed gourd on coarse textured sandy soils, particularly during drought due to low water content of the soil. However, the nematode also occurs in sandy clay loam and loam soils.

1.2 Parasitic nematodes incidence

Important diversity presence among the polyphagous nematodes of various localities of eastern part of India including a part of U.P. Mostly pointed gourd has been recorded as a host during frequent survey made by many scientists for at least one of the most frequently occurring species of root-knot nematode, *M.incognita*. Important other nematodes like *Hoplolaimus indicus, Rotylenchus reniformis, Tylenchorhynchus vulgaris* are only a local importance in crop growing areas (**Table 1**). Conversely

Nature of feeding	Common name	Scientific name	Symptom caused
A.Ecto-Parasitic	Lance nematodes	*Hoplolaimus indicus*	Stunting the vines and foliage of crop
	Spiral Nematodes	*Helicotylenchus dihystera*	
		H.abunaami	
		H. crenacauda	
	Dagger nematodes	*Xiphinema*	Root tip swelling
		Hirschmanniella gracilis	
		Criconemella ornata	
B.Endoparasitic	Root-knot nematode	*M.incognita*	Galls on roots,twigs or vines creeped on grounds
C.Semi-endoparasitic	Reniform nematodes	*Rotylenchulus reniformis*	Yellowing of foliage

Table 1.
Diversity in plant parasitic nematodes associated with pointed gourd.

Treatment	Fruit Yield (Q/ha)	Root-knot index(1–5)	Per cent loss in fruit yield
Carbofuran 2 kg a.i./ha	69.40 (+77.90)	1.00 (+80.00)	43.80
Untreated	39.00 (−43.80)	5.00	

Figures in parentheses show, per cent increase (+) or decrease (−) over untreated.

Table 2.
Yield losses due to M. incognita in pointed gourd (T.dioica Roxb.).

root-knot nematodes that predominate in upland region are in gangatic river belt of pointed gourd prone area [7, 9].

Root-knot nematode, which increase to damaging levels within a few season in susceptible crop are so common in perennial crop production that frequently they are taken to represent "Hidden enemy "in general. The other nematodes also cause heavy losses alone or in synergistically associated with other disease causing pathogens like fungi, bacteria, viruses etc.

1.3 Crop losses

No authentic information on crop loss due to attack of root-knot nematode in available. Thus an experiment was conducted at farmer field where crop was treated with Carbofuran@2.0 kg a.i./ha to determine the avoidable yield loss by keeping untreated check. Observation revealed that Carbofuran @2 kg a.i./ha reduced root-knot infection by 43.80% and which helped to increase the fruit yield by 43.80% over untreated control [3, 8]. It has been recorded drastic decline in marketable fruit yield when initial population of J_2 had 2–3 juveniles/gm. field soil which was above threshold level. The quantitative loss in fruit yield had 43.8 per cent (**Table 2**) where field was not protected with nematode but protected field crop had markedly higher fruit yield was observed through the fruit picking period with a seven day interval. In the non-protected plants, fruit yield is suppressed and difference was marked in months of July and August picking when fewer fruits have been picked but no such difference had noticed during March and April picking of crop season [10].

2. Symptomatology of root-knot nematode, *M.incognita* in crop

2.1 Symptoms

The common symptoms of root-knot nematode on pointed gourd have been found out the general stunting which are not grown as much as plants grown in nematode free soil, low vigour, chlorosis, necrosis, defoliation and twig die back. Twig galls has also been observed along with root galls (**Figure 2**) [9]. Infected plants are more susceptible to other diseases caused by fungi, bacteria [11] and tend to stop producing early. In pointed gourd the presence of galls on the root system and on propagated vines is the primary symptoms associated with Meloidogyne infection. During the warm days of July –August, the infected plants showed unhealthy growth and severe disease symptoms and a tendency to wilt. Stunting, non-emergence of sprouts, premature drying and shedding of leaves have been found in nematode infected fields where crop was being cultivated [2, 12].

DOI: http://dx.doi.org/10.5772/intechopen.98923

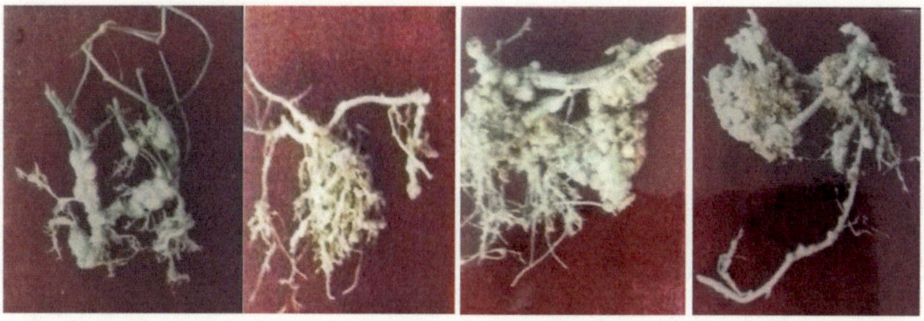

one year old crop Two year old crop Three year old crop Foure year old crop

Figure 2.
Profuse gall formation on vines of pointed gourd.

2.2 Gall formation

In galls formed by the nematode swelling of the central cylinder, highly deformed vascular elements and the spherical part of the nematode surrounded by the cortical parenchyma can be easily observed at low magnification in stained roots. During warm period, gall formation on roots and twigs is more conspicuous than in colder climate. Infected plants show fewer small rootlets, reduction in aerial growth in first year, while in second year and onwards crop showed marked decline in its production with the increase in nematode population and number of gall, smaller root system to support plant growth (**Figure 2**). The stunted plants showed poor root system, sometimes with large and confluent galls on the main root and twigs. The size and form of the galls depends on the species involved, number of nematode in the tissue host and plant age. In parwal the roots forms large, fleshy galls whereas twigs and shoot-galls unlike the root galls are of woody consistency. The size of galls varies considerably with age of plant parasitized by root-knot nematode species. In such cases the examination under the microscope revealed that infected young roots are full of pearly white nematode females attached by their heads and their egg masses covered by gelatinous matrix adhering with soil particles. This nematode completed its life cycle on pointed gourd within 30–45 days (**Table 3**) during warm season [13]. When plants are severely

Penetration and development stages(J_2)	Number, days after inoculation												
	1	3	6	9	12	15	18	20	24	27	30	34	P = 0.05
Penetration of J_2	145	118	182	140	—	—	—	—	—	—	—	—	3.72
Spiked tail stage	—	—	—	175	95	105	95	—	—	—	—	—	2.95
Moulting of J_2	—	90	75	70	105	130	—	—	—	—	—	—	1.87
Third stage	—	—	—	—	95	115	68	37	55	—	—	—	2.05
Fourth stage	—	—	—	—	—	105	80	75	44	26	40	—	2.14
Young female	—	—	—	—	—	—	40	60	50	85	45	—	1.98
Deposition of gelatinous matrix	—	—	—	—	—	—	—	130	45	65	20	—	1.65
Emergence of J_2	—	—	—	—	—	—	—	—	—	—	40	75	0.95

Penetration = 58.5%, eggs per egg sac = 385, male formed = 0.35%, females formed = 90.12%.

Table 3.
Biology of root-knot nematode, Meloidogyne incognita, on pointed gourd.

infected by *Meloidogyne* the normal root system is reduced to a limited number of severely galled roots with a completely disorganised vascular system. Rootlets are most completely absent. The translocation of nutrient and water by roots is severely hampered. Plants wilt rapidly, especially under upland growing conditions and are often stunted. Growth is retarded and leaves may be chlorotic. In case where infection at sprouting time has taken place, numerous plants die in the field and sprouting do not survive to grow new plants and creeping. The sprouts that survive to form new plants flowering and fruit production are highly reduced. As the season advances the galls are often invaded by fungi and bacteria that induce rotting. Wherever nematode populations are very high, young sprouts may be killed over large areas even without a trace of gall formation appearing on roots. In such cases the examination under the microscope will reveal that frequently the young roots are full of females attacked their heads and their egg masses covered by adhering soil particles [14, 15].

3. Biological study of root-knot nematode on crop

3.1 Biology and life cycle of nematode

It is a perennial crop, vegetatively propagated through vine cuttings and root suckers. One of the most important limiting factors in its profitable cultivation is heavy infestation by root-knot nematode. Due to non-availability of information on biology of nematode on this crop it was ascertained and determined the biology and reported that penetration of J_2 in roots continued up to 9 days with maximum numbers penetrating on 6th day. After penetration, the juveniles oriented themselves longitudinally near the vascular area behind the root tip and started moulting in 72 hrs. Young females appeared from 18th day after inoculation. Deposition of gelatinous matrix and egg-masses started from 20 to 24 days followed by emergence of J_2. Majority of the eggs were retained in the egg masses. The number of eggs varied from 50 to 385 per egg mass. The larval penetration in roots resulted in the formation of necrosis and irregular shaped syncytia. The infection also caused the formation of confluent round to spindle shaped galls laterally on roots [16] (**Table 3**).

The root-knot nematode is primarily root parasites. The adult females are sedentary and remain inside the root while males are vermiform and are inhabitant of soil. Sexual dimorphism is pronounced.

Several workers have been studied to determine the biology of root –knot nematode, *M.incognita* on pointed gourd. The J2 penetrated the root after 24 hour of inoculation and 58.5% penetration was recorded in roots while 90.12% penetrated J2 were moulted into different stage of juveniles however, moulting was started from third day and development of young females from 18th day after inoculation. The juvenile's stages (J3 and J4) become sedentary. Maturation of females was started from 20th to 22nd days. Deposition of gelatinous matrix and egg mass were started from 20th–24th days and emergence of J2 was initiated immediately even before the egg masses turned into brown. Whereas, number of males which were observed (0.35%) after 26th days of inoculation and eggs were 385 per egg mass. The fecundity of the nematode was not affected by the host. Thus root-knot nematode *M.incognita* is able to complete its life cycle from J2 to next generation within 30–35 days at a temperature ranged of 30-40°C(**Table 3**) [13, 17].

3.2 Ecology

The population density and damaging potential of root-knot nematode and other phytoparasitic nematode on pointed gourd vary considerably from field to field. During field survey conducted by [5, 7, 18] assayed the population density of each plant parasitic nematode and compared with growth parameter of pointed gourd. The strongest correlation between the population densities of root-knot nematode and growth responses recorded when soil assayed for nematodes were made on first, second and third year old crop. Root-knot nematode *Meloidogyne incognita* was the most damaging parasite as evidenced by high negative correlation between population densities and plant growth responses.

3.3 Pathogenicity

There is a study on which has been worked out the damaging threshold of *M.incognita* on pointed gourd and revealed that progressive decreased on plant growth was observed with increased in nematode inoculum level. Significant reduction in length, fresh weight of shoot and root was reported at different inoculum level (100, 1000, 10000 J_2) except at 10 J_2/plant. The number of galls, egg masses and multiplication of nematode continued to increase with the increasing inoculum level for pointed gourd [19].

4. Integrated management of root-knot nematode, *M.incognita* in pointed gourd crop

Reduction in crop yield due to nematode can be greatly managed by using available management practices [20]. Crop rotation is one of the oldest and most economic methods of controlling nematodes. However, these management practices must be taken before planting or propagating the crops through its vine nature. Once the nematode are persisted inside the roots/twigs (**Figure 2**) effective and potential treatments are not available, therefore, control strategies needs to be preventive rather than curative in nature and aimed from the onset at preventing the build-up of high population densities. Many techniques used for managing root–knot *Meloidogyne* invasion [21, 22] on pointed gourd simultaneously control other phytoparasitic nematode affecting the crop [23]. Combining and effective rotational scheme and selected cultural practices and use of chemical give excellent control with little added cost. In severely affected field, chemical/nematicides may be very useful [24] in lowering down the nematode population with its threshold level [25–27].

4.1 Cultural practices

Cultural practices may be minimised root-knot nematode damage. Practices such as removing the roots of each crop as soon as harvest is being completed, followed by tilling or summer deep ploughing of the soil two to three times is very effective in reducing nematode population [20].

4.1.1 Root-knot free field and propagating material

Nematode free planting material should be used for propagation. Field must be ensured free from root-knot, nematode in order to reduce dissemination. Chemical disinfection of propagating material is a common and effective measure in large areas where as other methods must be deployed for subsistence farming [28].

4.1.2 Crop rotation

Several workers [6, 11] have already been suggested rotation designed to reduce the impact of root-knot nematodes in tropical cropping system. A number of rotations exist in the pointed gourd growing areas which are predominantly composed of cruciferous crop, moderately resistant to tolerant against root-knot nematode. Usually farmers are grown pointed gourd perennially, it should be followed a mix cropping of inter-cropping system or companion cropping which can be reduced the susceptibility and promote tolerance of pointed gourd to root-knot nematode disease. Recommendations of a survey carried out by the various scientists [5, 18] helped to impede root-knot nematode invasion in pointed gourd by using the mustard crop in rotational cropping system.

4.1.3 Destruction of roots and vines

Practices such as removing the roots of each crop as soon as harvest is completed, followed by tilling the soil two to three times is every effective in reducing nematode levels. The tilling operations destroy the plant roots and prevent further reproduction of the nematode. It also exposes the nematodes to the drying action of the sun and wind, which reduces the level of nematode population. Maintaining optimum conditions for plant growth in terms of soil pH, fertility and soil moisture increases the tolerance of light to moderate nematode attack and makes plants less susceptible to other stresses as well. Galled roots and vines remaining in the field after harvest should be eliminated by uprooting and destruction. The spread of nematode can be retarded and the initial population density reduced because the nematode cannot survive and reproduce on the roots in the soil after harvest.

Practices such as removing the roots of each crop as soon as harvest is completed, followed by tilling the soil two to three times is very effective in reducing nematode levels. The tilling operations destroy the plant roots and prevent further reproduction of the nematode. It also exposes the nematodes to the drying action of the sun and wind, which reduces the level of nematode population. Maintaining optimum conditions for plant growth in terms of soil pH, fertility and soil moisture increases the tolerance of.

4.2 Organic amendments

In eastern U.P., India, growing areas of the crop wherein it was revealed the dreaded association of root-knot nematode, *M.incognita* infestation with unthrifty growth of vines in many areas (Verma and Anwar, 1993; Anwar, 2004). The root-knot nematodes delayed and suppressed the emergence and subsequent growth of sprouts, which had a marked influence on the performance of the host. The use of organic amendments in the soil is greatly emphasised as an alternative easy, cheap and satisfactory method of nematode control. The incorporation of chopped leaves of medicinal plants into the soil reduces root-knot densities [8, 21, 28]. The application of organic amendment as a means of biological control of root-knot disease leads to better plant growth and significant sprouting emergence, reduction in inoculum density and reduction in host susceptibility. Various oil cakes, different botanicals [3, 27, 29] and meals like neem (*Azadirachta indica*), mahuva (*Madhuca longifolia*) and castor(*Ricinus communis*) cake @ 25 qt/ha, plant parts of subabul (*Leucaena leucocephala*) and *Calotropis gigantea* are used as a source of organic amendments [29]. Further they have been revealed in their studies that powder form of cakes of mahuva (*M. longifolia)* and neem (*A. indica*) are incorporated/amended into the soil at the rate 250 kg/ha impeded the incidence of *M.incognita* in *T. dioica* but mustard cake application with the carbofuran applied at 2kga.i./ha have completely been suppressed the nematode population in soil and significantly avoid the invasion of J2 in roots of pointed gourd (**Table 4**). Although the use of organic amendments for effective nematode control is often limited by the large

Soil amendments with	Emerged sprouts per plant pit	Infected plants/ vines(%)	Length(cm) of vines	M.incognita population density in 200 cc soil	Galls/gm of root system	Females/ gm of root system	Root-knot index
Neem cake@25Q/ha	17.00	28.00	185.50	35.00	4.00	6.00	3.00
Mahuva cake@25Q/ha	12.00	32.00	172.300	42.00	5.00	8.00	4.00
Press mud @25Q/ha	13.00	49.00	175.10	50.00	6.00	14.00	4.00
Neem leaves @25Q/ha	18.00	30.00	181.60	40.00	3.00	7.00	4.00
Carbofuran @ 5 kg a.i./ha	14.00	24.00	187.40	32.00	3.00	6.00	3.00
Unamended field	6.00	100.00	145.50	200.00	20.00	25.00	5.00

Table 4.
Management of root-knot nematode, M.incognita *on pointed gourd (*T.dioica*).*

quantities needed, they will reduce nematode population densities to different level. In addition to their suppressive effect on nematode population they improve soil structure and water holding capacity [27]. Effect of plant extract or exudates and anthelmintic drugs on nematode hatching and mortality has been studied by different workers [23, 30] which indicated the natural occurrence of anti-nematode prohibitions in extract of different plant parts of marigold (*Tagetes erecta*). Their studies have been concluded and confirmed the toxic nature of tissue extract to M.incognita by inhibiting hatching process and increased mortality. The highest toxic effect had recorded in leaf extract followed by root, flower bud and seed extract. Mortality in nematode population gradually decreased when dilution of extracts increased in contrast to hatching.

4.3 Companion crops

Pointed gourd is an important profitable cash crop which is extensively cultivated in eastern part of Indian continent. This crop has been raised along with different marigold varieties as companion crops which have been shown lowed gall formation and egg mass development in roots. The highest toxic effect due secretion of metabolites and various root amino-acids on gall formation was exhibited by marigold varieties such as Saffron Spice variety of marigold allowed to lowest gall formation than that of Yellow Gate. Whereas, lowest reproduction factor has been recorded in root-knot nematode, *M.incognita* was observed when Saffron Spice variety of marigold was planted together and maximum had been observed in the companion planting of marigold varieties, FM-561 and Yellow Gate. The highest toxic effect of Saffron Spice has been recorded in reduction of soil population of *M.incognita, while, Hoplolaimus indicus, Helicotylenchus dihystera, Tylenchorhynchus vulgaris* and *Tylenchus* spp. were also to be reduced by FM-2,Hormony Boy and Sunset Geantee Variee [30].

4.4 Resistance sources

Resistant varieties reduce the population of root-knot nematode and produce a good crop even in the presence of nematodes. The effectiveness is increased when combined with crop rotation. By alternating root-knot resistant and susceptible crop within a given site from one year to the next, the overall nematode problem can be reduced by preventing a build-up of high populations. This practice may reduce the risk of serious damage to the susceptible crop. There are many reports of root-knot, Meloidogyne sp. parasitizing plants which have been reported non-host, an important factor in developing rotation based control system [8]. The effectiveness is increased when resistant host combined with crop rotation. By alternating resistant and susceptible crop within a given site from one year to the next, the overall nematode population can be reduced by minimising high population build-up. There is so any resistant material available in pointed gourd against root-knot nematode hitherto, Verma and Anwar, 1993 reported that except BP-2 all the germplasm of pointed gourd viz. BP-1,BP-3,BP-4,BP-5,BP-7 and BP-8 were highly susceptible to M.incognita. Variety BP-2 has some tolerance against root-knot nematode [5, 18].

4.5 Effect of potential Rhizospheric fungi

Various scientists have been revealed standard concentration of culture filtrate of saprophytes exhibited nematotoxic effect by inhibiting the hatching of root-knot nematode, *M. incognita*. Among the saprophytes, the minimum larvae had been encountered by *Aspergillus candidus* and *A. niger* while as *Verticillium albo- atrum* recorded maximum inhibitory effect on hatching [31]. Larval emergence had however inversely proportional to filtrate concentration which may be due to the differences in the nature of toxic metabolites by fungi. Species of Aspergillus [32], Penicillium and Trichoderma etc. are known to produce toxins and antibiotics like oxalic acid, malformin, penicillin and giotoxin [22]. Certain rhizospheres which are stimulated on addition of organic matter either become a part of mycoflora population for decomposing the organic matter or directly affect hatching of nematodes. It has revealed greater inhibition of root penetration by M.incognita, development of females and galling occurred with simultaneous presence of *R.solani* and *M. incognita* [33, 34]. They were confirmed with the in vitro trial wherein nematode inoculated prior and after, fungus exhibited slight to moderate inhibition of galling and female development but simultaneous presence of fungus and nematode showed linear decrease in final population of nematode in crop. Most of the fungi colonising eggs and egg masses in soil ecology where pointed crop was being cultivated [14]. Five genera of fungal fauna had been encountered from egg masses like *Fusarium(F.solani, F.oxysporum, F. dimerum); Aspergillus (A.flavus, A.niger, A. terreus); Drechslera(Drechslera sp., D. ravenelii); Rhizoctonia bataticola* and *Curvularia lunata* from crop prone areas which are exhibited toxic effect against hatching and survival of J_2 [14, 30, 33].

5. Chemical

Chemical control is the most advocated and no doubt paracticable method of root-knot nematode control. However, for pointed gourd crop of high economic value it becomes a must. The pit application of Carbofuran, phorate @ 2 kg a.i./ha gave a satisfactory control. Application of carbofuran, oncol and hostothian @ 0.1% as vine dip treatment are also useful in increasing sprouts and suppressing nematode population. The plants vines get infected when buried in soil fields for cultivating the crop fetching remarkable fruit yields of the crop. Various chemicals including different dosages of basamid Gr have been reported as successful in controlling by significant reduction in gall formation stimulating the growth enhancements of the crop [24].

DOI: http://dx.doi.org/10.5772/intechopen.98923

6. Conclusion

The crop fetches more prices in the market and its demand due to everyday consumption as vegetable and stupendous nutritive value. A preliminary survey of the crop in eastern U.P. India indicated the association of root knot nematode with unthrifty growth of plants in many areas. It is most susceptible to root-knot nematode resulting heavy loss of fruit yield reducing the income of marginal farmers whose are mostly cultivated in India and elsewhere. In case where nematode infection at sprouting time of pointed gourd has taken place, sprouting does not survive to grow new plants and its proper growth. The sprouts that survive to form new plants flowering and fruit production are highly reduced. As the season advances the nematode galls are often invaded by fungi and bacteria that induce rotting. Wherever nematode populations are very high, young sprouts may be killed over large areas even without a trace of gall formation appearing on roots. Reduction in crop yield due to nematode can be greatly managed by using available management practices. Crop rotation is one of the oldest and most economic methods of controlling nematodes. However, these management practices must be taken before planting or propagating the crops through its vine nature. Once the nematode are persisted inside the roots/twigs effective and potential treatments are not available, therefore, control strategies needs to be preventive rather than curative in nature and aimed from the onset at preventing the build-up of high population densities. Many techniques used for managing root–knot *Meloidogyne* invasion on pointed gourd simultaneously control other phytoparasitic nematode affecting the crop. Combining and effective rotational scheme and selected cultural practices and use of chemical give excellent control with little added cost. In severely affected field, chemical/nematicides may be very useful in lowering down the nematode population with its threshold level.

Author details

Ali Anwar*, Najeeb Mohammad Mughal, Efath Shahnaz, Saba Banday, Taibah Bashir, Qadrul Nisa and Gulam Jeelani
Faculty of Horticulture, Division of Plant Pathology, SKUAST-Kashmir, Shalimar, Srinagar, Jammu and Kashmir, India

*Address all correspondence to: zaman04@rediffmail.com; alianwar@skuastkashmir.ac.in

References

[1] Gopalan C, Ramasastri BV, Balasubramanian SC. Nutritive value of Indian Foods. I.C.M.R.:Hyderabad, India;1982. 80p.

[2] Verma A C, Anwar A. (1998). Nematode-A serious threat to pointed gourd (*Trichosanthes dioica*).In:Trivedi P C,editor. Nematode Diseases in Plants. CBS Publishers & Distributors,Daryaganj,New Delhi, India-110002;1998. P.361-368.

[3] Verma AC, Anwar A. (1995). Integrated nematode management by using leaves of medicinal plants in pointed gourd(T.dioica) field. Indian J.Nematol. 1995;25:5.

[4] Verma AC, Anwar A. Integrated management of root-knot nematode *M.incognita* on pointed gourd. Indian J.Mycol.Pl.Pathol. 1995;25:86.

[5] Verma AC,Anwar A. Second Annual Technical Report ICAR Ad-hoc Scheme Investigation on root-knot nematode on pointed gourd. Department of Nematology, NDAUAT, Faizabad, U.P. India; 1993.

[6] Page,SLJ. An assessment of the importance and control of plant-parasitic nematodes of vegetable crops in Bangladesh.O.D.M. Report of visit to Bangladesh.Ascot,Book,UK Ministry of Overseas Development. 1979;p.36

[7] Verma AC, Anwar A. Occurrence of root-knot nematode *Meloidogyne incognita* on pointed gourd(*Trichosanthes dioica* R.) in eastern UP., Indian J. Nematol. 1993;23:21.

[8] Verma AC, Anwar A. Integrated nematode management by using leaves of medicinal plants in pointed gourd (*Trichosanthes dioica*) field. Indian J. Nematol. 1994;25:5.

[9] Verma AC, Anwar A. Twig galls incited by *M.incognita* on *Trichosanthes dioica* R.. Indian J.Nematol. 1993;23 :215.

[10] Verma AC, Anwar A. (1996). Assessment of yield loss due to *Meloidogyne incognita* in pointed gourd,*Trichosanthes dioica* Roxb., Afro-Asian J. Nematol. 1996; 6:92-93.

[11] Sikora SA. Inter-relationship between plant health promoting rhizobacteria, plant parasitic nematodes and soil micro-organisms. Mededelingen van de Faculteit Landbouw wetenschappen, Rijksuni resiteit,Gent, Belgium.1988; 53:867-878.

[12] Mukherji SK, Sharma BD.1973. Root-knot disease of *Trichosanthes dioica*. Indian Phytopath. 1973 ;26:318-349.

[13] Verma AC, Anwar A.1994. Biological studies of root-knot nematode *Meloidogyne incognita* on pointed gourd (*Trichosanthes dioica* Roxb.) in eastern UP. In: Proceeding of 22nd International Nematology Symposium held at Gent, Belgium; 1994.p.72

[14] Prasad J, Anwar A, Verma AC. Fungi encountered with egg mass of root-knot nematode *M.incognita*. In: Proceeding of National Symposium on 'Rational approaches in nematode management for sustainable agriculture",Nematological Society of India,IARI,New Delhi; 1998.p.103-105

[15] Verma AC, Anwar A. (1998).Role of marigold (*Tagetes* sp.) as a companion crop against root-knot nematode, *M. incognita* in pointed gourd. In: Proceeding of National Symposium on'Rational approaches in nematode management for sustainable agriculture. Nematological Society of India,IARI,New Delhi; 1998.p.16-19

[16] Verma AC, Anwar A. (1999). Studies on Meloidogyne incognita on pointed gourd (*Trichosanthes dioica* Roxb.) in Eastern U.P., India. Indian J. Nematol. 1999; 29:190-91.

[17] Singh K. Pointed gourd (*Trichosanthes dioica* Roxb). Indian Horti.1987;34-37.

[18] Verma AC,Anwar A. 1992. First Annual Technical Report ICAR Ad-hoc Scheme Investigation on root-knot nematode on pointed gourd. Department of Nematology, NDAUAT,Faizabad,U.P.India; 1992.

[19] Verma AC, Anwar A. (1995). Pathogenicity of *Meloidogyne incognita* on pointed gourd (*Trichosanthes dioica* Roxb.).Indian J.Mycol.Pl.Pathol. 1995;25:70.

[20] Anwar A, Verma AC. 1994. How to control root-knot nematode in parwal. Farm Digest(NDUAT Supplement)1994:25.

[21] Anwar A. Effect of organic amendment on interaction between wilt disease caused by *Fusarium oxysporum f.sp.lycopersici* and root-knot nematode, *M.incognita* in tomato crop. In: Proceeding of International Seminar on Recent Trend Hi-Tech Hort. & PHT,Kanpur,U.P.India,2004. P.198-199

[22] Man kau, R.Phytopathology, 1969; 59:1170-1172.

[23] Mohammad HY,Husain SI, Al-Zarari AJ. Effect of plant extracts of some poisonous plants of Iraq on the mortality of citrus nematode, *Tylenchulus semipenitrans* Cobb. Acta. Bot.Indica.1981;9:198-200.

[24] Verma AC, Anwar A. Efficacy of Basamid Gr. Against *M. incognita* in tomato nursery. Afro-Asian J. Nematol. 1994;4:73-75.

[25] Nath RP,Haider MG, Akhter SW, Prasad H. Studied on the nematode of vegetable in Bihar I. Effect of reniform nematode, *Rotylechulus reniformis* on *Trichosanthes dioica*. Indian J. Nematol. 1976: 6:175:77.

[26] Ray S, Das SN. Nematode Fauna of Orissa Technical Bull. Published by Dept.Nematology OUAT, Bhubneswar,Orissa; 1989.114p

[27] Verma AC, Anwar A. (1997). Control of *Meloidogyne incognita* on pointed gourd. Nematol. Medit. 1997;25:31-32.

[28] Verma AC, Anwar A. (1998). Effect of organic amendments on sprout emergence of pointed gourd, in root –knot nematode, *M.incognita* infested field. Ann.Pl.Protec.Sci. 1998; 6:102-104.

[29] Anwar A, Khan FU. Effect of aqueous leaf extracts of medicinal plants on the growth of rhizospheric fungi of tomato cv.Pusa Ruby in vitro. SKUAST J. Research. 2001; 3(1):60-63.

[30] Verma AC, Anwar A. (2000).Herbal effect of marigold tagetes varieties on hatching and mortality of root-knot nematode, *Meloidogyne incognita*. In: Proceeding of Indian Phytopathology Golden Jublee;2000. P.691-692.

[31] Anwar A. Effect of rhizospheric fungi of tomato cv. Pusa Ruby on the hatching of root-knot nematode, *Meloidogyne incognita*. Agic.Sci. Digest.2004; 24:59-60.

[32] Anwar A, Saxena SK. Effect of culture filtrate of *Aspergillus niger* Van Tiegh on growth of tomato plants and development of *Rotylenchulus reniformis* linford and Olivera, 1940. Current Nematol.1993; 4(2):207-210.

[33] Anwar A, Verma AC. Interaction between *M. javanica* and *Rhizoctonia solani* on chickpea, *Cicer arietinum* L. Ann. Pl. Protec. Sci. 1993;1:137-138.

[34] Anwar A, Khan FA,Saxena SK. Interaction between *Meloidogyne incognita* and *Rhizoctonia solani* on soyabean. In: Proceeding of the Second Afro-Asian Nematology Symposium, Menoufiya, Egypt, 1996. P.110-113

Chapter 6

Effects of Irrigation and Bioproducts of Microbial Origin on Nematode Community and Mycorrhizal Root Colonization in Soybean

Ivana Majić, Ankica Sarajlić, Emilija Raspudić, Marko Josipović and Gabriella Kanižai Šarić

Abstract

Soybean (*Glycine max* L. Merr) is the most important legume and threaten by diverse pests and diseases. Complex interactions among rhizosphere organisms are found in all agro-ecosystems. Results of these interactions can be positive and/or negative in terms of plant production. Soil nematode community consists of different trophic groups of nematodes. Nematodes are the most abundant soil invertebrates. Several nematode species penetrate soybean roots as parasites, and can cause loss in yields. Arbuscular mycorrhiza fungi are obligate plant symbionts that colonize soybean roots naturally. The aim of the study was to evaluate effects of irrigation and amendments of bioproducts containing beneficial soil microorganisms (ABM) on nematode community and mycorrhizal root colonization in soybean. Field experiments were conducted in soybean in 2013 in Osijek, Croatia. The plots were either rain fed or irrigated to 60-100% field water capacity (FWC). We tested soil amendments and soil + foliar amendments of three commercial products containing beneficial organisms. Average number of nematodes per soil sample varied from 186,67 (soil ABM in non-irrigated plots) to 297,57 (soil+foliar ABM in plots with 60-100% FWC), and there were no significant differences between the treatments. Bacterial feeding nematodes were the most abundant, while plant parasitic genus *Pratylenchus* was the most abundant among other plant parasitic nematodes. There was no clear influence of any of the treatments on soil nematode community. Amendments of the bioproducts increased mycorrhizal root colonization in rain fed plots, while it decreased the mycorrhizal root colonization when soybeans were irrigated. Irrigation increased mycorrhizal root colonization in plots without amendments of the bioproducts, and mycorrhizal colonization differed significantly between the sampling dates. Further research is needed to determine if irrigation alters the potential of mycorrhiza to colonize the roots.

Keywords: soybean, nematodes, *Pratylenchus*, arbuscular mycorrhiza fungi, irrigation, soil and foliar amendments, beneficial microorganisms

1. Introduction

Soybean (*Glycine max* L. Merr) is economically the most important legume. The largest production area under soybean is in North and South America (USA, Brasil, Argentina), China and India [1]. Soybean is used for food and feed because of the rich nutrition profile, proteins and oil. For that reason, it is widely used in different industry branches such as food, oil, pharmaceutical, textile and chemical industries [2]. An abiotic stress is a major constraint in crop production. Drought can reduce soybean yield over 50% annually. Drought stress is also transmitted from parental plants to F1 generation and reduces the seed germination rate, therefore optimal water supply is a must for the best seed quality [3]. In temperate climatic regions where natural precipitation during the growing season is lower than 300 mm, irrigation is necessary [4]. Adaptation of soybean to abiotic conditions in a site, along with plant interactions with other living organisms, represents principal factors for successful crop production [5]. Inoculation of soybean plants with beneficial bacteria and mycorrhizal fungi can facilitate water stress and increase yield, as much as other parameters like seed fat content [6].

Rhizosphere or soil near the soybean root zone is the most dynamic environment of microbe-plant interaction [7]. Soil organisms depend on each other for carbon and energy, and represent major component for assessment of soil health. Several groups of organisms are distinguished in rhizosphere, mainly saprophytes and plant symbionts. Multi trophic interactions in soil directly influence the biodiversity of soil organisms and indirectly promote plant growth and ability to withstand pathogen attack [8].

Plant growth-promoting microorganisms (PGPM) include bacteria, and fungi, that live in soil and rhizosphere and stimulate plant growth by synthesizing phytohormones, producing siderophores, fixing atmospheric nitrogen, dissolving inorganic forms of elements such as phosphorus, and increasing plant resistance to stress and abiotic biotic environmental conditions [9–11]. The most commonly used inoculant of PGPM in the soybean crop belongs to rhizobium bacteria that colonize the root creating nodules which supplies plant with biologically fixed atmospheric nitrogen. Mixed cultures of microorganisms such as *Bradyrhizobium* with *Azospirillum*, *Bacillus*, *Pseudomonas* and *Glomus* are considered valuable and used in soybean production [10, 12, 13]. This type of co-inoculation shows great efficiency especially in soils where stressful environmental conditions such as low phosphorus content prevail [10]. The use of mineral fertilizers can be minimized when soybeans are inoculated with PGPM, and this measure is desirable since it is environmentally sustainable [13]. Higa and Parr [14]. isolated group of beneficial microorganisms from the soil and named them effective microorganisms. This group included approx. 80 species, mostly photosynthetic bacteria, lactic acid bacteria, yeasts, actinomycetes, and fermenting fungi such as *Aspergillus* and *Penicillium*.

2. Interactions of nematodes and mycorrhizal fungi in rhizosphere

Nematodes are the most abundant soil invertebrates, with beneficial and detrimental role in agriculture. They serve as good bioindicators of the effect of agricultural practices and contaminants on the functioning of the soil food web [15]. Mainly, five nematode trophic groups are commonly found in agricultural soils: plant parasitic, bacterial feeding, fungal feeding, omnivorous and predaceous [16]. Soybean is an important oilseed crop and source of high quality protein. Plant parasitic nematodes are economically important plant pathogens for all agricultural plants, and pose treat to production of globally demanding high yield soybeans. The most important soybean nematodes are soybean cyst nematodes (*Heterodera glycines* Ichinohe) and root lesion nematodes (*Pratylenchus* spp.). In Croatia, *Pratylenchus* is the most frequent and abundant genus of plant parasitic nematodes found in soybean, however economically important yield reductions due to root lesion nematodes damage have not been reported [17].

Rhizosphere microorganisms are often antagonistic to plant parasitic nematodes. By decomposing organic matter in rhizosphere, microorganisms release nematicidal compounds in surrounding soil. Their derivates are often toxic and negatively affect nematodes. Soil microorganisms also compete with nematodes for the same source of food, or feed upon the nematodes. Lowering the amount of space for living and food source, the microorganisms could suppress nematode population. This interaction occurs in both directions. Nematophagous fungi (e.g. *Paecilomyces* spp., *Pochonia* spp., *Verticilium* spp., *Trichoderma* spp. etc.) and antagonistic bacteria (eg. *Pasteuria penetrans*, *Pseudomonas fluorescens* etc.) are soil-borne microorganisms that are very useful bioagents against plant parasitic nematodes.

Efficacy of biocontrol agents often depends on ability to adopt to different cropping techniques and soil conditions. Bioproducts containing beneficial microorganisms are mostly registered as fertilizers or plant growth promoters, and claim to enhance plant tolerance and defense system, and finally increase yields by suppressing plant parasitic nematodes by associating with mycorrhiza [18]. Interactions among beneficial soil organisms and plant parasitic nematodes are mainly evaluated under laboratory or greenhouse conditions [19]. Nematode trophic groups other than plant parasitic are beneficial, since they contribute to nitrogen mineralization by feeding on and by dispersing beneficial bacteria, also they are regulating rates of decomposition [20].

Arbuscular mycorrhiza fungi (AMF) are obligate symbionts that colonize the roots of most cultivated plant species. The most plant species form mycorrhizal symbiosis naturally [21]. Association of plants with AMF increase the absorptive surface of the plant root system, enhance plant access to immobile soil minerals, and increase plant growth rates, respectively. Mycorrhizal symbiosis provides soybean with nutrients, mitigates abiotic stress such as draught and improves host plant resistance against pests and diseases [21]. It induces a variety of physiological and molecular biological changes in the host plant and may improve plant resistance and tolerance to the most important plant parasitic nematodes [22]. Direct and indirect effects of AMF on rhizosphere organisms are observed as results of altered plant exudation Direct and indirect effects on the soil biota may include altered plant exudation, and via competition and mutualism [23]. The aims of the study were to evaluate effects of irrigation and amendments of beneficial soil microorganisms nematode community and mycorrhizal root colonization in soybean.

3. Experimental design

Field experiments were conducted in soybean in 2013, at Agricultural Institute Osijek, Croatia (45°32" N and 18°44″ E, altitude 90 m). Size of the experimental field was 405 m^2. The field has a history of a long-term soybean-maize rotation. The soil is characterized as eutrical non-calcareous brown soil developed on calcareous loess substrate middle gleyed and silt/clay loam texture. Soybean (cultivar Ika) was grown and maintained by conventional farming practices. To examine effects of amendments of bioproduct of microbial origin on soil nematodes and mycorrhizal colonization, plots were assigned to three types of treatments: control, soil amendments and soil + foliar amendments in irrigated and rain fed plots. The experiment was set according to randomized block design in three replicates. Three commercial bioproducts were used in experiment: EM Aktiv (Multikraft), Nourivit and Nourivit plus (Nourivit Technologies GmbH). These products contain more than 40 different species of beneficial soil microorganisms (mainly lactic acid bacteria, photosynthetic bacteria, and yeasts) and sugarcane mollases, claimed by the manufacturer. In plots with soil amendments, EM Aktiv was applied in dosage 30 L ha^{-1} prior sawing of soybean to enable activation of microorganisms. In plots with soil + foliar treatment, soil amendment of EM Aktiv (30 L ha^{-1}) was applied

Figure 1.
The experimental field setup: a) soybeans in the experimental plots, b) the watermark sensors of soil moisture, c) self-propelled sprinkler (typhon), d) boom irrigation systems.

Treatment	Quantities of the irrigated water (mm) and date of the treatments		Grain yields (kg ha^{-1})
	mm	date	
Control	0	—	3000
Irrigated plots	35	June 3-6	4050
	35	July 12-14	
	35	July 21-23	
	35	August 2-4	

Treatments: control – 0 mm of added water by irrigation; irrigated plots – maintenance of soil water content from 60 to 100% field water capacity (FWC).

Table 1.
Irrigation schedule at the experimental site.

prior sawing and two foliar treatments of Nourivit (4,5 kg ha^{-1}) and Nourivit plus (4,5 L ha^{-1}) were applied during vegetation.

Self-propelled sprinkler (typhon) was used to irrigate plots 60–100% of the field water capacity (FWC) (**Figure 1**). Irrigation depended of the soil water content, the weather characteristics, mainly precipitations (**Table 1**).

Quantities of water and frequency of irrigation are presented in **Table 1**. Irrigation rate was 35 mm, and following measures of soil moisture were done at root depth of 30 cm. The method consisted of Watermark sensors and hand-held field meter. Sensors were buried in the soil after sowing the soybeans at two depths: 15-20 cm and 25-30 cm and removed after the harvest. Measurements were taken twice a week or after the significant rainfall and irrigation regime.

Long-term mean (LTM, 1961-1990) of precipitations is 368 mm during the growing season (April–September) in Osijek (**Figure 2**). Investigated area has a semi-humid and drought prone climate. In 2013, the environmental conditions were moderate, with minimal deviations from total precipitation and air temperature of LTM.

Soil and root sampling for nematode and mycorrhizal fungi analysis was done twice during the vegetation, in July and September. Extraction of nematodes from soil was done following modified Baermann funnel method [24]. Nematodes were counted and separated according to their feeding habit to trophic groups [16]. and according to the morphological characteristics plant parasitic nematodes were identified to the genus level [25, 26]. Ten soybean roots were excavated in three replications from each and subjected to microscopic analysis for mycorrhizal colonization. Soybean root and mycorrhizal preparation, and staining was done according to the method described by Vierheilig et al. [27]. The presence of mycorrhizae was determined according to the method described by Trouvelot et al. [28]. and following parameters were determined: mycorrhizal frequency in the root system (F), intensity of mycorrhizal colonization in the root system (M), arbuscule abundance in the root system (A), intensity of mycorrhizal colonization in the root fragments (m) and arbuscule abundance in mycorrhizal parts of root fragments (a). The data were log(n+1) transformed prior analysis of variance (PROC GLM). The means are back-transformed and presented in Tables. The means were separated by Tukey test (P<0,05) (SAS 9.2; SAS Institute, Carey, NC, USA).

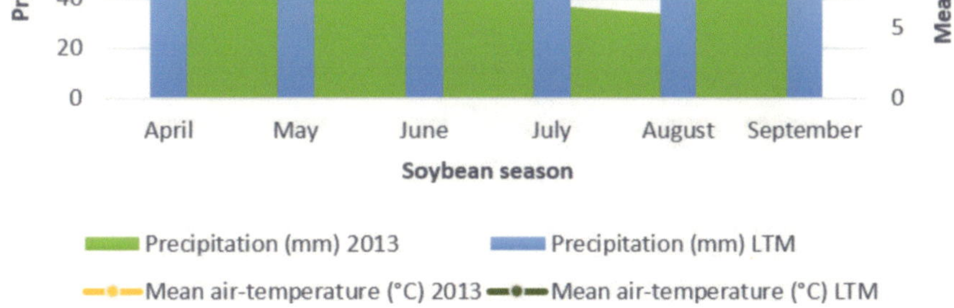

Figure 2.
Precipitation and air-temperature in growing season 2013 and long-term means (LTM) (Osijek weather bureau).

4. Results

4.1 Influence of irrigation and amendments of beneficial organisms on soil nematode community

Soil samples in soybean were taken twice to estimate the effect of irrigation and amendments of bioproducts of microbial origin on nematode trophic groups (**Table 2**). According to F-statistics for group of plant parasitic nematodes irrigation level (F=11,51, P<0,001), month of sampling (F=14,95, P<0,001) and interaction treatment*month was found as statistically significant (F=5,56, P<0,001). Population density of bacterial feeding nematodes significantly changed only when affected simultaneously by two variables treatment*month (F=4,58, P<0,05). Omnivorous nematodes were significantly affected only by the month of sampling (F=4,16, P<0,05), while significant response of predators was found only in simultaneous effect of two variables irrigation*month (F=14,14, P<0,05). Group of fungal feeding nematode did not respond significantly to any of the tested variables. Statistics revealed that interaction of treatment with bioproducts and month of sampling significantly affects total nematode community (F=3,47, P<0,05).

	Nematode trophic group					Total
	PP	F	B	O	P	
Irrigation	11,51**	0,95	1,91	0,38	0,45	1,84
Treatment	1,78	0,66	1,68	1,95	1,20	1,51
Month	14,95**	0,03	1,37	4,16*	0,45	0,85
Irrigation*Treatment	0,00	0,64	0,04	1,45	2,49	0,02
Irrigation*Month	0,31	0,42	0,14	1,52	14,14*	0,03
Treatment*Month	5,56**	0,44	4,58*	0,46	2,49	3,47*

Data are F-values; PP – plant parasitic, F – fungal feeding, B – bacterial feeding, O – omnivorous, P – predator.
**P<0,05.*
***P<0,001.*

Table 2.
GLM analysis of effect of irrigation level and amendment of bioproducts on nematode trophic groups densities.

Plant parasitic nematodes	Effects		
	Treatment	Month	Treatment*Month
Ditylenchus	1,71	9,27**	1,41
Filenchus	0,32	4,29*	1,25
Malenchus	1,63	0,20	3,20*
Merlinius	0,60	2,16	0,17
Pratylenchus	3,58*	3,44	1,87
Tylenchorynchus	0,83	4,36*	1,72
Tylenchus	0,28	3,68	1,65

Data are F-values.
**P<0,05.*
***P = 0,001.*

Table 3.
GLM analysis of the effects of treatments with bioproducts, month of sampling, and their interaction on plant parasitic nematodes in soybean.

Seven plant parasitic nematode genera were identified from soil samples of each treatment (**Table 2**). Treatment of soybean with bioproducts significantly affected plant parasitic genus *Pratylenchus* ($F=3,58$, $P<0,05$) (**Table 3**). Month of sampling significantly affected population of *Ditylenchus* ($F=9,27$, $P<0,001$), *Filenchus* ($F=4,29$, $P<0,05$), and *Tylenchorynchus* ($F=4,36$, $P<0,05$). Simultaneous effect of treatment and month of sampling was statistically significant for the population of *Malenchus* ($F=3,20$, $P<0,05$).

Nematodes belonging to the genus *Pratylenchus* were the most abundant among all other plant parasitic nematodes (**Table 4**). In the treatment with a soil amendment of bioproducts, populations of *Pratylenchus* were significantly higher compared to the treatment with two amendments of bioproducts (i.e. soil+foliar). The lowest population density of *Pratylenchus* spp. (31,86% of plant parasitic nematodes per soil sample) was found in plots with soil and foliar amendment. However, the treatments did not significantly differ from the control plots, where on average 42,94% *Pratylenchus* spp. of total plant parasitic nematodes per soil sample was identified. In previous studies from Croatia, *Pratylenchus* was also the most dominant genera in soybean [17, 24, 29, 30]. In Brasil, one of the world's leading soybean production area plant parasitic nematodes are major constrain and the most dominant nematode trophic group [31]. In the same study, *Pratylenchus*, *Helicotylenchus* and *Meloidogyne* were found as the most important plant parasitic nematode genera in soybeans. Total population of plant parasitic nematodes in our study did not significantly differ when comparing types of amendments of bioproducts.

Another study tested long term amendments of effective microorganisms, compost and mineral fertilizers on soil nematode community [32]. The results of the cited study showed that effective microorganisms applied together with compost increased the abundance of total bacterial and plant parasitic nematodes compared to the plots with mineral fertilizer, compost and control. Plant parasitic nematodes were the most dominant trophic groups in their study, and increased in relative abundance by 34.33% in effective microorganisms' plots compared to the mineral fertilizer plots. Wheat biomass in the cited study was also increased by amendments of effective microorganisms, which could be the reason for increase in plant parasitic nematodes populations, since more food was available. Amendments of manure, a source rich with diverse species of microorganisms, to soil increase abundance of nematode community [33].

Plant parasitic nematodes	Treatments		
	Control	Soil	Soil+foliar
Ditylenchus	1,86 a	2,92 a	5,20 a
Filenchus	16,04 a	8,33 a	19,36 a
Malenchus	7,29 a	0,83 a	2,50 a
Merlinius	1,25 a	0,90 a	2,91 a
Pratylenchus	42,94 ab	71,67 a	31,86 b
Tylenchorynchus	9,38 a	20,08 a	16,25 a
Tylenchus	8,96 a	11,25 a	11,67 a

Data are percentage of relative nematode abundance; Values in rows with different letters are statistically significant at P<0,05.

Table 4.
Analysis of variance for the effects of amendments of bioproducts on plant parasitic nematodes.

4.2 Influence of irrigation and amendments of beneficial organisms on mycorrhizal root colonization

Statistical analysis revealed significant influence of irrigation (F=34,95, P<0,001), month of sampling (F=94,70, P<0,001), and interaction irrigation*treatment and treatment*month (F=14,29, P<0,001; F=13,16, P<0,001) on mycorrhizal frequency in the root system (**Table 5**). Intensity of mycorrhizal colonization in the root system and in root fragments is under a significant influence of irrigation (F=38,17, P<0,001; F=34,16, P<0,001), treatment (F=4,48, P<0,05; F=6,00, P<0,05), month (F=145,99, P<0,001; F=61,87, P<0,001), interaction irrigation*treatment (F=17,29, P<0,001; F=13,40, P<0,001) and treatment*month (F=11,43, P<0,001; F=16,78, P<0,001). Arbuscule abundance in the root system and in root fragments was significantly affected by irrigation (F=10,99, P<0,01; F=21,92, P<0,001) and all interactions: irrigation*treatment (F=14,73, P<0,001; F= 6,04, P<0,05), irrigation*month (F=68,83, P<0,001; F=95,42, P<0,001) and treatment*month (F=10,11; P<0,001; F=3,33, P<0,05).

Effects of commercial AMF products on growth, nutritional, and physiological responses of soybean in another study reveal the difference between the products with regard to their response to water deficit [34]. Inoculation of plants with AMF was found more important than soil moisture in improving plant growth to overcome drought stress [35]. We found month of sampling and irrigation as the most important factor for mycorrhizal root colonization. However, treatments with bioproducts were similarly important only in interaction with date of sampling.

4.3 The importance of irrigation on the effect of different amendments of beneficial organisms

In previous study, irrigation and nitrogen fertilization increased significantly soybean grain yields [36]. The grain yields in this study were also considerably increased in irrigated plots with more than 1000 kg ha^{-1} difference between control and irrigated plots (**Table 1**). Average number of nematodes per soil sample varied from 186,67 (soil treatment in non-irrigated plots) to 297,57 (soil+foliar treatment in plots with 60-100% FWC), and there were no significant differences between

	Root colonization (%)				
	F	**M**	**A**	**a**	**m**
Irrigation	34,95***	38,17***	10,99**	21,92***	34,16***
Treatment	2,06	4,48*	2,56	0,99	6,00*
Month	94,70***	145,99***	0,08	1,17	61,87***
Irrigation*Treatment	14,29***	17,29***	14,73***	6,04*	13,40***
Irrigation*Month	1,09	0,70	68,83***	95,42***	3,45
Treatment*Month	13,16***	11,43***	10,11***	3,33*	16,78***

Data are F-values; F – mycorrhizal frequency in the root system, M – intensity of mycorrhizal colonization in the root system, A – arbuscule abundance in the root system, a – arbuscule abundance in mycorrhizal parts of root fragments, m – intensity of the mycorrhizal colonization in the root fragments.
**P<0,05.*
***P<0,01.*
****P<0,001.*

Table 5.
GLM analysis of effects of treatments with bioproducts, month of sampling, irrigation level, and their interaction on mycorrhizal colonization of root system.

		Irrigation level					
		Control			60-100% FWC		
Bioproduct amendment		Soil	Soil+foliar	Control	Soil	Soil+foliar	Control
Nematode community	PP	7,92 a	6,14a	6,32a	11,87b	9,23a	9,65a
	B	152,50a	212,50a	206,25a	235,00 a	265,00 a	155,83a
	F	15,42a	28,95a	16,25 a	22,50 a	20,00a	19,58a
	O	9,17a	10,00a	12,50a	10,00 a	3,33a	3,33a
	P	1,67a	1,60a	0a	2,50a	0a	2,50a
Total		186,67a	259,27a	241,32a	281,87a	297,57a	190,90a
Mycorrhizal root colonization	F	25,78b	30,11b	16,28a	18,55a	6,89a	20,78b
	M	3,31b	3,93b	2,09a	1,99b	0,79a	4,62b
	A	0,37b	0,22ab	0,17a	0,02a	0,03a	0,39b
	m	11,70b	9,52ab	7,61a	5,23a	3,32a	12,66b
	a	11,05a	6,33a	7,23a	1,06a	1,23a	3,72a

Data are means of nematode population density and percentage of mycorrhizal root colonization; Values in rows marked with different letters are statistically significant at P<0,05; PP – plant parasitic, F – fungal feeding, B – bacterial feeding, O – omnivorous, P – predator; F – mycorrhizal frequency in the root system, M – intensity of mycorrhizal colonization in the root system, A – arbuscule abundance in the root system, m – intensity of the mycorrhizal colonization in the root fragments, a – arbuscule abundance in mycorrhizal parts of root fragments.

Table 6.
The effects of different amendments of bioproducts of microbial origin and irrigation level on nematode community and mycorrhizal root colonization.

the treatments (**Table 6**). Bacterial feeding nematodes were the most abundant in all treatments, ranging from 152,50 to 265 nematodes per sample, but no significant effects of treatments were found as well. Soil amendment of bioproduct in irrigated plots significantly increased the number of plant parasitic nematodes, where in average 11,87 nematodes were found per sample. Nematodes live in a film of water around the soil particles, so they respond quickly to any changes in environments and irrigation could affect the nematode survival. In another study, only the proportion of omnivores and the number of taxa identified was affected by irrigation [37]. Artificial irrigation could change soil physical and chemical properties, and abundance and diversity of nematode community is correlated with these changes [38]. In this study, there was no clear influence of irrigation on soil nematode community.

Mycorrhizal root colonization frequency in the root system was higher in the plots with bioproduct amendments in non-irrigated plots (**Table 6**). Significantly highest mycorrhizal frequency in the root system (30,11%) was observed in the treatment with soil+ foliar amendments of bioproduct in non-irrigated plot. This treatment had the greatest impact on root mycorrhiza. Irrigation also significantly affected the mycorrhizal colonization, since F was as low as 16.28% in non-irrigated control (without bioproduct amendments), compared to significantly high F (20,78%) in irrigated control. However, irrigation affected the mycorrhizal colonization in plots with bioproducts amendment. Bioproduct increased mycorrhizal root colonization for all tested parameters in non-irrigated plots. When applied twice in soybean vegetation, in soil and on foliar, bioproduct significantly decreased the mycorrhizal colonization (for all parameters, except parameter a) in irrigated regime 60-100% FWC. In other studies, higher mycorrhizal root colonization was observed during the dry comparing to the wet period [39]. but different results were also reported [40]. Difference in mycorrhizal root colonization also depend on the plant genotype [24].

5. Concluding remarks

Studies aiming to evaluate effect of commercial products containing beneficial microorganisms on soil nematode community, especially on plant parasitic nematodes and mycorrhizal root colonization are scarce. Nematode community structure respond quickly to changes in their environment resulting from agricultural practices and other changes in soil properties. The results we presented in this chapter reveal weak effect of irrigation and amendments of bioproducts containing beneficial organisms on abundance and structure of nematode community in soybean. However, the treatments we used had considerable effect on mycorrhizal root colonization. This result is positive for soybean production, since AMF could increase plant performance in drought stress and consequently impact on greater grain yields. Our results indicate that amendments of the bioproduct increase mycorrhizal root colonization in rain fed plots, while it decreased the mycorrhizal root colonization when soybeans were irrigated.

Acknowledgements

This work was partially financed by the project "Biological control of the European Corn Borer (Ostrinia nubilalis Hübner)" (Ministry of Science, Education and Sports, Croatia; Grant no. 079-0790570-2208). Many thanks to Croatian Waters, Zagreb for financial support of the project "Irrigation, soil and water protection in sustainable agriculture in Eastern Croatia."

Conflict of interest

The authors declare no conflict of interest.

Author details

Ivana Majić[1]*, Ankica Sarajlić[1], Emilija Raspudić[1], Marko Josipović[2] and Gabriella Kanižai Šarić[1]

1 Faculty of Agrobiotechnical Science Osijek, University of Osijek, Osijek, Croatia

2 Agricultural Institute Osijek, Osijek, Croatia

*Address all correspondence to: imajic@fazos.hr

References

[1] FAO. Crops [Internet]. 2021. Available from: http://www.fao.org/faostat/en/#data/QC [Accessed: 2021-6-16]

[2] Sudarić, A. Soybean for Human Consumption and Animal Feed. London, United Kingdom. IntechOpen; 2021. DOI:10.5772/intechopen.73719.

[3] Wijewardana C, Reddy KR, Krutz LJ, Gao W, Bellaloui N. Drought stress has transgenerational effects on soybean seed germination and seedling vigor. Plos one. 2019; 14(9): e0214977. DOI:10.1371/journal.pone.0214977

[4] Gajić B, Kresović B, Tapanarova A, Životić LJ, Todorović M. Effect of irrigation regime on yield, harvest index and water productivity of soybean grown under different precipitation conditions in a temperate environment. Agricultural Water Management. 2018; 210: 224-231 DOI:10.1016/j.agwat.2018.08.002.

[5] Bello LL, Shaahu A, Vange T. Studies on relationship between seed yield and yield components in soybean (Glycine max L. Merrill). Electronic journal of Plant Breeding 2012; 3 (4): 1012-1017.

[6] Igiehon NO, Babalola OO, Cheseto X, Torto B. Effects of rhizobia and arbuscular mycorrhizal fungi on yield, size distribution and fatty acid of soybean seeds grown under drought stress. Microbiological Research. 2021; 242: 126640. DOI:10.1016/j.micres.2020.126640

[7] Hiltner L, Über neuere Erfahrungen und Probleme auf dem Gebiet der Bodenbakteriologie und unter besonderer Berücksichtigung der Gründüngung und Brache, Arb. Dtsch. Landwirtsch. Ges. 98 (1904) 59-78.

[8] Lynch JM, The Rhizosphere, Wiley, New York, USA, 1990; p. 458.

[9] Mishra J, Singh R, Arora NK. Plant growth-promoting microbes: Diverse roles in agriculture and environmental sustainability. In Kumar V, Kumar M, Sharma S, Prasad R, editors. Probiotics and Plant Health. Singapore: Springer; 2017. p. 71-111. DOI:10.1007%2F978-981-10-3473-2_4

[10] Santo MS, Nogueira MA, Hungria M. Microbial inoculants: Reviewing the past, discussing the present and previewing an outstanding future for the use of beneficial bacteria in agriculture. AMB Expres. 2019; 9:205. DOI:10.1186/s13568-019-0932-0

[11] Bakhshandeh E, Gholamhosseini M, Yaghoubian Y, Pirdashti H. Plant growth promoting microorganisms can improve germination, seedling growth and potassium uptake of soybean under drought and salt stress. Plant Growth Regulation. 2020; 90: 123-136. DOI:10.1007%2Fs10725-019-00556-5

[12] Zeffa DM, Fantin LH, Koltun A, de Oliveira ALM, Nunes MPBA, Canteri MG, Gonçalves LSA. Effects of plant growth-promoting rhizobacteria on co-inoculation with Bradyrhizobium in soybean crop: A meta-analysis of studies from 1987 to 2018. Peer J. 2020; e7905. DOI:10.7717/peerj.7905

[13] Meng L, Zhang A, Wang F, Han X, Wang D, Li S. Arbuscular mycorrhizal fungi and rhizobium facilitate nitrogen uptake and transfer in soybean/maize intercropping system. Frontiers in Plant Science. 2015; 6: 339. DOI:10.3389/fpls.2015.00339

[14] Higa T, and Parr JF. Beneficial and effective microorganisms for a sustainable agriculture and environment. Atami: International Nature Farming Research Center. 1994; 1-16.

[15] Neher DA, and Campbell CL. Sampling for regional monitoring of

nematode communities in agricultural soils. Journal of Nematology. 1996; 28: 196-208.

[16] Yeates GW, Bongers T, De Goede RGM, Freckman DW, Georgieva SS. Feeding habits in nematode families and genera – An outline for soil ecologists. The Journal of Nematology. 1993; 25(3): 315-331.

[17] Majić I. Endoparazitne nematode roda Pratylenchus na soji. Poljoprivreda, 2010; 16(2): 57-58.

[18] Drobek M, Frąc M, Cybulska J. Plant biostimulants: Importance of the quality and yield of horticultural crops and the improvement of plant tolerance to abiotic stress—A review. Agronomy. 2019; 9(6): 335.

[19] Dong LQ, and Zhang KQ. "Microbial control of plant-parasitic nematodes: A five-party interaction. Plant and Soil. 2006: 288 (1): 31-45.

[20] Neher DA. "Nematode communities as ecological indicators of agroecosystem health. Agroecosystem sustainability: Developing practical strategies. 2001; 105-120.

[21] Smith SE, and Read DJ. Mycorrhizal symbiosis, Acad. Press, London, UK, 1997.

[22] Li HY, Yang GD, Shu HR, Yang YT, Ye BX, Nishida I, Zheng CC. Colonization by the arbuscular mycorrhizal fungus Glomus versiforme induces a defense response against the root-knot nematode Meloidogyne incognita in the grapevine (Vitis amurensis Rupr.), which includes transcriptional activation of the class III chitinase gene VCH3. Plant and Cell Physiology. 2006; 47(1): 154-163.

[23] Tibbett M. Roots, foraging and the exploitation of soil nutrient patches, the roles of mycorrhizal symbiosis. Functional Ecology. 2000; 14: 397-399

[24] Majić I, Ivezić M, Raspudić E, Gantner V. Effect of soybean cultivar on endoparasitic nematodes and arbuscular mycorrhizal fungi relationship. Cereal Research Communications. 2008; 36: 1823-1826.

[25] Bongers T. De Nematoden van Nederland. KNNV-bibliotheekuitgave 46. Pirola, Schoorl. 1994; p. 408

[26] Mai WF, Mullin PG, Lyon HH, Loeffler K. Plant-Parasitic Nematodes: A Pictorial Key to Genera. Edition 5. Cornell University Press. 1996; p. 288

[27] Vierheilig H, Coughlan AP, Wyss U, Piché Y. Ink and vinegar, a simple staining technique for arbuscular mycorrhizal fungi. Applied and Environmental Microbiology. 1998; 64(12): 5004-5007. DOI:10.1128/AEM.64.12.5004-5007.1998.

[28] Trouvelot A, Kough JL, Gianinazzi-Pearson V. Mesure du taux de mycorhization VA d'un système radiculaire. Recherches de méthodes d'estimation ayant une signification fonctionnelle. In: Gianinazzi-Pearson V, Gianinazzi S, editors. Physiological and Genetical Aspects of Mycorrhizae. Paris: INRA; 1986; 217-221.

[29] Ivezić M, Majić I, Raspudić E, Brmež M. Occurrence of soil and plant nematodes in soybean under cereal rotation. Cereal Research Communications. 2008; 36: 431-434.

[30] Raspudic E, Ivezic M, Samota D. *Pratylenchus* species of soybean in Croatia 1. EPPO Bulletin. 1994; 24(2): 399-402.

[31] Gomes SG, Huang SP, Cares JE. Nematode community, trophic structure and population fluctuation in soybean fields. Fitopatologia Brasileira. 2003; 28(3): 258-266.

[32] Hu C, Qi Y. Effective microorganisms and compost favor

nematodes in wheat crops. Agronomy for Sustainable Development, Springer Verlag/EDP Sciences/INRA. 2013; 33(3): 573-579. DOI:10.1007/s13593-012-0130-9

[33] Villenave C, Saj S, Pablo AL, Sall S, Djigal D, Chotte JL, Bonzi M. Influence of long-term organic and mineral fertilization on soil nematofauna when growing Sorghum bicolor in Burkina Faso. Biology and Fertility of Soils. 2010; 46(7): 659-670.

[34] Al-Karaki GN, Williams M. Mycorrhizal mixtures affect the growth, nutrition, and physiological responses of soybean to water deficit. Acta Physiologiae Plantarum. 2021; 43: 75. DOI:10.1007/s11738-021-03250-0

[35] Shukla A, Kumar A, Jha A, Salunkhe O, Vyas D. Soil moisture levels affect mycorrhization during early stages of development of agroforestry plants. Biology and Fertility of Soils. 2013; 49:545-554

[36] Josipović M, Sudarić A, Kovačević V, Marković M, Plavšić H, Liović I. Irrigation and nitrogen fertilization influences on soybean varieties (*glycine max* (L.) Merr.) properties. Poljoprivreda/Agriculture. 2011; 1: 9-15.

[37] Zhi D, Li H, Nan W. Nematode communities in the artificially vegetated belt with or without irrigation in the Tengger Desert, China. European Journal of Soil Biology. 2008; 44(2): 238-246.

[38] Fiscus DA, Neher DA. Distinguishing sensitivity of free-living soil nematode genera to physical and chemical disturbances, Ecological Applications. 2002; 12(2): 565-575. DOI:10.1890/1051-0761(2002)012[0565,DSOFLS]2.0.CO;2

[39] Birhane E, Sterck FJ, Fetene M, Bongers F, Kuyper TW. Arbuscular mycorrhizal fungi enhance photosynthesis, water use efficiency, and growth of frankincense seedlings under pulsed water availability conditions. Oecologia. 2012; 169: 895-904. DOI:10.1007/s00442-012-2258-3

[40] Bhardwaj AK, Chandra KK. Soil moisture fluctuation influences AMF root colonization and spore population in tree species planted in degraded entisol soil. International Journal of Biosciences. 2018; 13: 229-243. DOI:10.12692/ijb/13.3.229-243

Chapter 7

Root-Knot Nematodes a Major Peril to Protected Cultivation System in India: Current Status and its Management

Jaydeep A. Patil and Saroj Yadav

Abstract

Growing of vegetable crops under protected conditions are relatively, an innovative technology and most popular among farmers throughout the country. In last few decades protected cultivation has shown potential enhancement in horticultural production. The southern root-knot nematode, *Meloidogyne incognita*, is an emerging nematode under protected conditions. This nematode can cause chlorosis, stunting and reduce yields associated with the induction of many root galls on host plants. Root-knot nematode severely affect the plant root system by inducing specialized feeding cells i.e., giant cells in the vascular tissues. Recently, this nematode has been considered as a worldwide menace for combat root-knot nematodes, integrated nematode management strategies such as soil solarization, biological control, organic amendment, crop rotation, field sanitation, and fumigants have been developed and successfully used in the past. Here, in this book chapter discussed on biology and life cycle, control measures and proposed future strategies to improve *Megalaima incognita* management under protected conditions.

Keywords: protected cultivation, root knot nematode, vegetable crop and integrated nematode management

1. Introduction

Cultivation of crops in protected structures is relatively a new or advance technology, growing crops in controlled environments (temperature, humidity, light and such other factors can be regulated as per requirement of the crop). It is popular among farmers/growers globally. Commonly used structures are forced ventilated greenhouse, naturally ventilated polyhouse, high-tech polyhouse, insect proof net house, shade net house, plastic tunnel and mulching. Protected structures may be demarcated as "Alteration of environmental condition in such a way to accomplish maximum growth and yield" [1]. Recently, incipient technology for raising high value crop in the country and it has very decent potential in semi-urban areas (nearby cities). Altered environmental conditions, bounces manifolds increase in yield per unit area. Modernized protected cultivation are very popular among growers in all over world and approximately 405000 ha area covered under protected cultivation globally [2] as compared to India 30000 ha area under protected cultivation, is still in infancy stage [3]. In India, protected structures are being initiated by National Horticulture Mission to increase per capita income of framers. Protection of pest and diseases under controlled environmental conditions, farmers are getting very good returns from this technology. Globally, among polyhouse cultivated crops, *Cucumis sativus* L. is an important vegetable and second most popular crop.

In polyhouses, three types of crops are grown, *viz*. vegetable crops such as cucumber, capsicum, tomato, ornamental crops such as, carnation, roses, gerbera, chrysanthemum and fruit crop like strawberry. Growing vegetables and flower crops under protected cultivation is receiving utmost attention and gaining popularity among farming community across the country. The ideal conditions provided by protected cultivation and continuous availability of the host plant round the year often results in high population buildup of soil borne pathogens including plant parasitic nematodes. However, Plant parasitic nematodes are becoming a major constraint in production of the horticultural crops under protected cultivation in India. Root-knot nematode, *Meloidogyne* spp. has to be the major plant parasitic nematode under protected conditions [4]. There are various management strategies viz., soil solarization, biological control, organic amendment, chemical and integrated nematode management practices have been followed for the management of the plant parasitic nematodes.

Protected structures aided crops with altered climatic conditions to get supreme yield potential than open field by shielded from adversities [5]. Ancient records, during 14–37 AD, when Roman Empire was controlled agricultural production, certain limited structures were present. Nevertheless, commercial protected cultivation had been initiated in England trailed by France, Netherlands, Japan and China at ending of eighteenth and nineteenth century [6]. Charles Lucien Bonaparte, French botanist (1803–1857) are accredited for making the first modern greenhouse (http://english.reachgreenhouse. com/news_view_32_105.html). High value agricultural crops are mostly preferred for protected structures to optimize production cost as well as reduced biotic and abiotic stresses.

2. Prevalence of root-knot nematode under protected cultivation

Root-knot nematode, *Meloidogyne* spp. are foremost important parasite in protected cultivation and having ability to parasitize on most of the crops. Around 232 plant species attacked by root-knot nematodes including vegetables, fruits, fiber, ornamentals, medicinal, cereals and weeds also. Root-knot nematodes are obligate parasites causes severe damage to vegetable crops (**Figures 1–3**), leading to major yield reductions and significant economic losses worldwide [4, 7–9]. Root-knot nematode development and fecundity are very high and very with populations/ races. Second stage juveniles randomly move in soil and attracted by chemicals released from host roots thus seek to infect roots. Meristematic zone is the most preferred site for penetration of second stage juveniles. Galls on roots are the diagnostic symptom, results from hypertrophy and hyperplasia after nematode feeding. As endoparasites, all stages were found in root tissues, except vermiform male and second stage. Optimum temperature required for development was 15 to 30°C. One generation, from egg to egg, was completed within 25 to 30 days. In a conducive environmental condition, nematode population build-up increases rapidly and reaches as high as 20–25 eggs and juveniles per g soil.

Root galls are the most characteristic symptom of root-knot nematode infection (**Figure 1**). As vascular feeder, destroy the xylem and phloem, ultimately translocation of water and nutrient uptake was debilitated. Due to poor transportation, above ground symptoms such as yellowing, wilting, poor fruiting has manifested on plants in patches. These patches gradually increase every year as inoculum has increased.

Figure 1.
Cucumber crop infested wish root-knot nematode, Meloidogyne spp. under polyhouse conditions. A) above ground symptom B) below ground symptoms. (Source: Original photos).

3. Interaction of *Meloidogyne* spp. with other microorganisms

Nematode have long been alleged of playing a bigger role in plant disease scenario with other microorganisms like fungus and bacteria rather than alone. Different role has been played by nematode in complex diseases such as aggravator, predisposer, vector etc. nematodes may act as wounder, host substrate modifier, rhizosphere modifier and vector to make the environment more conducive for development of other secondary pathogens. In complex diseases, root-knot nematode with wilt and root rot fungus causes grater damage to susceptible plants as compared to each pathogen alone [10, 11]. Host physiology has been altered by nematode for secondary pathogens results, complete failure of crop in some instances.

Nematodes provide ready avenues for entry of secondary microbes. Besides avenues, biochemical changes have been initiated in nematode infected plants and enriched giant cells also favor the wilt causing fungi. Seedling mortality was

Figure 2.
Tomato crop infested wish root-knot nematode, Meloidogyne spp. under polyhouse conditions. A) above ground symptom B) below ground symptoms. (Source: Original photos).

preponed by about a week due to interaction with fungi with nematode. Nematode may also play the role in resistance braker for other pathogens. Some varieties lost the resistance against fungal pathogens in the presence of root-knot nematodes. Due to disease complexes sometimes complete crop failure faces by growers.

4. Reasons for multiplication of root-knot nematode under protected cultivation

4.1 Moisture

Moisture is the foremost important factor for multiplication of nematodes. Continuous moisture availability around root zones through drip irrigation under

(a)

(b)

Figure 3.
Capsicum (Bel pepper) crop infested wish root-knot nematode, Meloidogyne spp. under polyhouse conditions A) above ground symptom B) below ground symptoms. (Source: Original photos).

polyhouses is responsible for fast nematode build-up and movement as compared to open field where flooded and dry conditions prevail. Continuous moisture availability flare-up the nematode population and more infectious.

4.2 Temperature

Temperature affects overall life profile activities of nematode such as hatching, movement, reproduction, development, and survival and also the host plant. Optimum temperature range for survival of plant parasitic nematodes is 15-30°C and become inactive or less active from above and below temperature conditions or may lethal for nematode. Under polyhouses optimum range of temperature for nematode build-up exist. Under optimum moisture and temperature conditions in polyhouses nematodes are able to complete several generations in less period of time as compared to open field conditions.

5. Continuous cultivation of susceptible host

Crops grown under polyhouses are generally susceptible to nematode pest such as tomato, cucumber, ornamentals etc. due to high economic value monoculture has been adopted by growers. Intensive monoculture of susceptible crops leads to nematode build-up and multiplication rate. All three conditions, susceptible host with favorable microclimate favor fast build-up of nematode population and once it introduced it is very difficult to get rid from this.

6. Current management approaches

Favorable climatic conditions concentrated majority of the protected structures in the regions between 25° and 65° latitude [12]. Solar irradiations and temperature is low at higher latitude, that's wants maintenance of humidity and temperature and the conditions created favor the pest incidence. Intense solar irradiation at lower altitude persuades stress in the crops rendering them susceptible to pest incidence [12]. Irrespective of the diverse protected structures and materials for buildup, the microclimate inside the protected structures favor the multiplication of pest and diseases including plant parasitic nematodes [13]. Henceforth, it become extremely difficult to manage the nematode pest from single management options. Keeping in view of this, integration of all available management techniques/tools for better resolution of the nematode pest. The integrated strategies for control of plant parasitic nematodes can be based upon two basic principles: 1) preventive measures and 2) on-farm techniques. Preventive measures avoid the introduction pest species in newer areas and second one is based upon control measure (cultural, biological and chemical) adopted by growers, to reduce pest population below ETL.

7. Preventive measures

Preventive measures are adopted to avoid the introduction of nematode pest in newer areas where nematode problem not exist before. Some practices have been adopted as preventive measures to control the spread of nematode. New or emerging species spread has been checked by regulatory methods to avoid the introduction in newer areas. Soil testing are mandatory for all the farmers before erection of polyhouses, green houses and net houses for plant parasitic nematodes. Entry points for protected conditions should also contain sanitizing stations for hands, shoes, boots, tools, and other equipment. Nematode free transplanting material is one of the important methods to avoid the nematode infestation under field conditions. Always use nematode-free transplants or plants that build upon soilless substrates from production are increasingly used to exclude soil borne species of nematodes, but also to promote the plant establishment and crop production.

Raising of crop on soilless media: One of the most important method to prevent spreading of nematodes in nematode free areas through growing of nursery crops in soilless media such as organic growing media: peat, coir, bark, sawdust, compost; inorganic: rockwool, perlite, pumice, sand, vermiculite.

8. Curative measures

Curative measures are used to reduce the nematode population below economic threshold level in nematode infested areas so, growers can get maximum returns.

Sanitation can minimize the nematode problems from polyhouses include rapid destruction of infested plant debris and weeds after harvest.

8.1 Soil solarization

Soil solarization is a most effective method to reduce the nematode population in hot weather areas (temperature around 40-50° c). In India, northern conditions are best to adopt this practice to reduce nematode infestation under polyhouses. Transparent polyethylene plastic (25 μm thick LLDP) mulch is used to cover the moist soil for 4–8 weeks in the month of May–June [14–17]. Green house effects have been created under transparent polyethylene sheet leading to higher temperature was lethal to nematode.

8.2 Crop rotation and inter cropping

Cultural practices are non-chemical method such as crop rotation with resistant cultivars or non-host crops to reduce pest population. Rotating or inter cropping tomato/ cucumber with non-hosts such as garlic (*Allium sativum*), Marigold (*Tagetes* sp.) (**Figures 4** and **5**), lettuce, radish, cabbage and cauliflower could reduce root knot nematode populations in soil. Few options are available in protected conditions to grow non-host crop, so, resistant cultivars are a very good option under protected structures.

8.3 Resistant cultivars

Resistant cultivars are one of the convenient options against plant parasitic nematodes. Grafting of commercially desired susceptible cultivars on resistant rootstock is a trending method among vegetable crops under protected conditions [18]. Resistant rootstock of brinjal wild relatives, *Solanum toxicarium*, *Solanum sisymbriifolium* and *S. torvum* have been grafted by commercial tomatoes, noticeable reduction in galling was observed [19]. Various grafted rootstock of melon and capsicum were produced that confirmed extraordinary results in minimize root galling in the greenhouses [20].

Figure 4.
Cucumber intercrop with marigold for the management of root-knot nematode under polyhouse conditions.

Figure 5.
Marigold crop rotation with cucumber/tomato for management of root-knot under polyhouse conditions.

8.4 Organic amendments

Enormous organic amendments are used for suppression of plant parasitic nematodes in protected cultivation. Suppression efficacy of organic amendments depends on the active ingredient and their concentration. Non-edible oil cakes of Neem (*Azadirachta indica*), castor (*Ricinus communis*) Karanj (*Pongamia glabra*), Mahua (*Madhuca latifolia*) etc. are used for management of rot-knot nematode in protected cultivation [21]. Other organic substances like FYM, vermicompost, slurry, green manure etc. are also effective for suppression of PPNs.

8.5 Biological control

Higher efficiency, targeted results, environmentally sound and local acceptability among the growers gain much popularity of the bio-agents in recent era. Egg parasitic fungus- *Paecilomyces lilacinus, Pochonia clamydosporia,* antagonistic fungus- *Trichoderma viride, T. harzianum,* VAM fungus- *Glomus* spp., bacterial parasite- *Pasteuria penetrans* and PGPR bacteria- *Pseudomonas fluorescence* is used as potential bio-agents against plant parasitic nematodes [17, 22–24]. Bio-agents enriched organic amendments are very effective strategy to control nematodes in protected cultivation.

8.6 Chemical nematicides

Till now, there are not a single nematicide registered for protected cultivation use in India. Thus, the growers depend on other integrated pest management practices for nematode management under polyhouses. Combination of all preventive, curative measures to control nematode under polyhouses is an effective strategy and locally adopted by growers.

9. Novel methods of resistance to root knot nematode under protected conditions

Wide susceptibility range, fast multiplication and cause potential treat at low density are the main constraints for management of root-knot nematode under protected structures. Recently, genetic engineering has made it possible to express and incorporate heterologous and indigenous protein from one to other organisms and develop heightened pest resistance in plants. Genetic engineering approaches has made natural resistance with synthetic resistance may be the auspicious tools for management of nematode in tomato production [25–27].

RNA Interference (RNAi): RNAi has emerging tool to downregulate gene activity and recognized efficient tactic against root-knot nematode [28]. RNAi first performed for *Caenorhabditis elegans* and it was used for gene silencing by overwhelming their expression in a plant parasitic nematode [29]. Nematode feeding site formation gene has been silenced by using dsRNA or siRNA that elicit a systemic RNAi response [30]. Root-knot nematode produces effector proteins determined by parasitism genes, and these effectors epitomize the molecular interface between the nematode and host [28]. Effectors secreted in nematode esophageal glands play perilous roles in parasitism [31].

Exploiting Efficient Genome Editing Using the CRISPR-Cas9 Technique: The advancement of the clustered regularly interspaced short palindromic repeats (CRISPR) technology has become a commanding alternative to gene silencing [32]. Foreign DNA sequences has incorporated host loci to produce short crRNAs (CRISPR RNAs) that direct sequence-specific cleavage of homologous target dsDNA by Cas endonucleases [33]. Recently, documentation of pathogen and host novel genes responsible for infection help in developing the CRISPR technique for improving the resistance to Meloidogyne spp. under protected systems.

Advantages of protected cultivation

- Higher productivity and higher income
- Quality produce
- Off season or round the year cultivation
- Hardening of tissue culture plants
- Better management of insect pest
- Less use of chemicals
- Efficient use of resources

10. Downsides of polyhouse cultivation

In spite of protected structures crops grown under these structures are not fully protected from insect pest. Hostile environment, intensive or mono cropping, availability of moisture (drip irrigation) and poor hygienic conditions are increasing the pest problems mainly nematodes under protected environment [34]. Among plant parasitic nematodes root-knot nematodes is the important parasite under polyhouses [4]. Once nematode introduced in the protected cultivation, it's impossible to eradicate the nematode problem. It can build up in less time and causes huge number of losses among the crops. Major source of adulterations in protected structures are planting material, soil and pooting media, water and general cleanliness.

11. Conclusions and future directions

Recently, nematode and soil borne pathogens under protected structures paid much attention. Till now efficient management practices under protected structures are very less and are not uses due to certain limitations. Researchers has focused on the environmentally sound conventional and modern management practices under protected structures. Biological and ecological aspects are the fundamental science to manage nematodes.

Author details

Jaydeep A. Patil* and Saroj Yadav
Department of Nematology, College of Agriculture, Chaudhary Charan Singh Haryana Agricultural University, Hisar, Haryana, India

*Address all correspondence to: rajhau99@gmail.com

References

[1] Singh, B. (2005). Protected cultivation of vegetable crops. Kalyani Publisher; pp. 168.

[2] Reddy, P.P. (2016). Sustainable crop protection under protected cultivation. Springer Singapore. https://doi.org/10.1007/978-981-287-952-3.

[3] Shweta, Bhatia, S.K., Malik, M. (2014). Protected Farming. Popular Kheti, 2: 74-79.

[4] Patil, J., Kumar, Anil, Goel, S. R. (2017a) Incidence of Plant-Parasitic Nematodes Associated with Polyhouses under Protected Cultivated in Haryana. Environment and Ecology, 35 (3A): 1870-1873.

[5] Maynard, E. and O'Donnell, M. (2018). Managing the environment in high tunnels for cool season vegetable production. https://mdc.itap.purdue.edu/item.asp?Item_Number=HO-297-W.

[6] Wittwer, S.H., Castilla, N., 1995. Protected cultivation of horticultural crops worldwide.

[7] Mekete T, Decraemer W, Wesemael WML, Seid A and Fininsa C. (2015). Tomato (*Solanum lycopersicum*) and root-knot nematodes (*Meloidogyne* spp.) – a century-old battle. Nematology. 17:1-15.

[8] Moens M, Viaene N. and Wesemael WML. (2011). Root-knot nematodes (*Meloidogyne* spp.) in Europe. Nematology. 13:3-16.

[9] Talavera M, Sayadi S, Chirosa-Ríos M, Salmerón T, Flor-Peregrín E and Verdejo-Lucas S. (2012). Perception of the impact of root-knot nematode-induced diseases in horticultural protected crops of south-eastern Spain. Nematology, 14:517-527.

[10] Agrios, G.N. 1988. Plant pathology. Sydney, Australia, Academic Press.

[11] Patil, J., Goel, S.R. and Yadav, S. (2018a) Effect of *Meloidogyne incognita* and *Fusarium oxysporum* f. sp. *cucumerinum* on cucumber grown under protected cultivation. Journal of Entomology and Zoology Studies, 6 (1): 1004-1007.

[12] Berlinger, M.J., Jarvis, W.R., Jewett, T.J., Lebiush-Mordechi, S. (1999). Managing the greenhouse, crop and crop environment. In: Albajes, R., Gullino, M.L., van Lenteren, J.C., Elad, Y. (Eds.), Integrated Pest and Disease Management in Greenhouse Crops. Kluwer Academic Publishers, Netherlands, 97-123.

[13] Heinz, K.M., van Driesche, R.G., Parella, M.P., 2004. Biocontrol in Protected Culture. Ball Hort Technology, 5: 6-23.

[14] Katan, J. (2017a) Diseases caused by soilborne pathogens: biology, management and challenges. J. Plant Pathol. 99 (2), 305-315.

[15] Katan, J. (2017b). Diseases caused by soilborne pathogens: biology, management and challenges. Journal of Plant Pathology, 99: 305-315.

[16] Kumar, A., Patil, J.A, Verma, K.K. (2019). Management of root-knot nematode, *Meloidogyne* spp. in vegetable crops grown under protected cultivation through fumigants. *Indian Journal of Nematology*, 49 (2): 125-130

[17] Patil J, Kumar A, Yadav S, Goel SR (2018b) Nematicidal effect of fumigants on the *Meloidogyne incognita* and *Fusarium oxysporum* f. sp. *cucumerinum* on cucumber in polyhouse. *Plant Pathol.* Journal, 17 (1): 25-32.

[18] Rivard, C.L., Sydorovych, O., O'Connell, S., Peet, M.M. and Louws,

F.J. (2010). An economic analysis of two grafted tomato transplant production systems in the United States. Horticulture Technology, 20: 794-803.

[19] Black, L.L., Wu, D.L., Wans, J.F., Kalb, T., Abbass, D., Chen, J.H. (2003). Grafting Tomato for Production of in Hot–Wet Season. AVRDC Publication, pp. 1-6: 03-551.

[20] Kokalis-Burelle, N and Rosskopf, E.N. (2011). Microplot evaluation of rootstocks for control of *Meloidogyne incognita* on grafted tomato, muskmelon, and watermelon. Journal of Nematology. 43: 166-171.

[21] Patil, J.A., Yadav, S. and Kumar, A. (2020). Evaluation of organic oils for the management of root-knot nematode, *Meloidogyne incognita* and Fungus infesting cucumber under polyhouse conditions. Indian Journal of Nematology, 50 (2) 79-86.

[22] Patil, J., Goel, S.R. and Yadav, S. (2017b) Bio-management of cucumber wilt complex caused by root-knot nematode, *Meloidogyne incognita* and *Fusarium oxysporum* f. sp. *cucumerinum* in polyhouse under protected cultivation. *Journal of Pure and Applied Microbiology*, 11 (4): 1909-1917.

[23] Patil J, Kumar A, Yadav S, Goel SR, Bhatia AK (2018c) Bio-Efficacy of Phyto therapeutic substances against *Meloidogyne incognita* and *Fusarium oxysporum* f. sp. *cucumerinum* affecting cucumber in polyhouse under protected cultivation. Indian Journal Nematogy, 48 (2): 190-197.

[24] Patil, J.A., Yadav, S. and Kumar, A. (2021). Management of root-knot nematode, *Meloidogyne incognita* and soil borne fungus, *Fusarium oxysporum* in cucumber using three bioagents under polyhouse conditions. *Saudi Journal of Biological Sciences*, https://doi.org/10.1016/j.sjbs.2021.07.081. Publishing, USA.

[25] Gheysen, G., van der Eycken, W., Barthels, N., Karimi, M., and Van Montagu, M. (1996). The exploitation of nematode-responsive plant genes in novel nematode control methods. *Pesticide Science,* 47: 95-101. doi: 10.1002/(sici)1096-9063(199605)47:1<95::aid-ps390>3.0.co;2-i.

[26] Jung, C., Cai, D., and Kleine, M. (1998). Engineering nematode resistance in crop species. Trends Plant Science, 3: 266-271. doi: 10.1016/s1360-1385(98)01247-3.

[27] Opperman, C. H., Acedo, G. N., Saravitz, D. M., Skantar, A. M., Song, W., Taylor, C. G., et al. (1998). "Bioengineering resistance to sedentary endoparasitic nematodes," in Advances in Molecular Plant Nematology, eds F. Lamberti, C. de Giorgi, and D. M. Bird (New York, NY: Plenum Press), 221-232. doi: 10.1007/ 978-1-4757-9080-1_19.

[28] Elling, A. A. (2013). Major emerging problems with minor *Meloidogyne* Species. Phytopathology, 103:1092-1102. doi: 10.1094/phyto-01-13-0019-rvw.

[29] Ali, M. A., Azeem, F., Abbas, A., Joyia, F. A., Li, H., and Dababat, A. A. (2017). Transgenic strategies for enhancement of nematode resistance in plants. Frontiers in Plant Science, 8:750. doi: 10.3389/fpls.2017.00750.

[30] Lilley, C. J., Davies, L. J., and Urwin, P. E. (2012). RNA interference in plant parasitic nematodes: a summary of the current status. Parasitology, 139: 630– 640. doi: 10.1017/s0031182011002071.

[31] Haegeman, A., Mantelin, S., Jones, J. T., and Gheysen, G. (2012). Functional roles of effectors of plant-parasitic nematodes. Gene 492, 19-31.

[32] Ali, M. A., Shahzadi, M., Zahoor, A., Dababat, A. A., Toktay, H., Bakhsh, A., et al. (2019). Resistance to cereal

cyst nematodes in wheat and barley: an emphasis on classical and modern approaches. International Journal of Molecular Sciences, 20:432. doi: 10.3390/ ijms20020432.

[33] Jinek, M., Chylinski, K., Fonfara, I., Hauer, M., Doudna, J. A. and Charpentier, E. (2012). A programmable dual-RNA-guided DNA endonuclease in adaptive bacterial immunity. Science, 337: 816-821. doi: 10.1126/science.1225829.

[34] Minuto, A., Gullino, M.L., Lamberti, F. D., Adabbo, T., Tescari, E. and Garibaldi, A. H. (2006). Application of an emulsifiable mixture of 1, 3 Dichloropropene and chloropicrin against root-knot nematode and soil fungi for greenhouse tomato in Italy. Crop Protection, 25, 1244-1252.

Section 3

Biological Control of Plant-Parasitic Nematodes

Chapter 8

Molecular Characterization and Pathogenicity of *Trichoderma* Isolates to *Meloidogyne javanica*

Ricardo R. Balardin, Cristiano Bellé, Daiane Dalla Nora, Rodrigo F. Ramos, José Carlos V. Rodrigues and Zaida I. Antoniolli

Abstract

Nematodes are considered a serious problem for agriculture. Nematodes of the *Meloidogyne* genus can attack a wide range of plants, needing different management methods to decrease its population. Fungi from the *Trichoderma* genus has been related to have potential as biological control agents. However, before an organism is used as biological control agent, first it is necessary to prospect, characterize and test its potential as biocontrol agent, so the objective of this work was to characterize and test fungi isolates of the *Trichoderma* genus to control *M. javanica*. We obtained forty isolate to carry out this experiment. We extracted the DNA of each isolate to discover which species we were testing, by doing a PCR and sequencing. We tested *in vitro* their parasitism effect using ELISA plate. Also, we extracted their filtrate to see if their metabolites have potential to reduce nematode population by showing a high mortality or inhibiting hatching. The results confirmed the high potential of the fungi of *Trichoderma* genus as a biological agent to control *Meloidogyne javanica*.

Keywords: Biological control, Integrated management, root-knot nematode, fungal filtrates, nematicide

1. Introduction

Soybean (*Glycine max* L.) is one of the most important agricultural commodities in the world. For this reason, it is constantly sought to increase productivity without increasing the cultivated area. Yield can be affected by several factors, among them are diseases caused by plant-parasitic nematodes, an important cause of reduced grain production in this crop. Worldwide, there are approximately 100 species of known plant-parasitic nematodes that decrease the production of this commodity [1]. In Brazil, these species are distributed primarily among the genera *Meloidogyne, Heterodera, Pratylenchus* and *Rotylenchus* [1–4]. Among these species, the root-knot nematodes, specifically *Meloidogyne javanica*, stands out because it has the ability to parasitize and cause significant damage to soybeans [5, 6].

Agricultural systems have had little crop diversification over the years, which means that these organisms have good availability of food throughout the year.

This availability causes the nematode population to grow more and more, making control even more difficult. Therefore, new control alternatives are being studied to minimize the damage caused by these organisms. Currently, biological control within integrated management stands out as an efficient and economically viable alternative to the use of chemical nematicides [7, 8]. In general, biological products have low toxicity and environmental risk, and we would be using a wide variety of microorganisms that can naturally parasite nematodes and their eggs in natural and agricultural environments. Among the main groups of microorganisms responsible for the biocontrol of nematodes, fungi stand out, representing up to 75% of the microorganisms used in the control of plant-parasitic nematodes [9, 10]. The ability to colonize the soil and persist for a long period makes these organisms increasingly visible. When we think about long persistence, we are comparing it with chemicals, which do not have a long-lasting residual. We can observe in the field chemical products that are used in seed treatment in soybean culture lose their efficacy even before the reproductive period of the culture.

Therefore, the possibility of incorporating organisms that have a long persistence in the soil can be a very important control measure, given the worrying scenario that the nematodes have been presenting. Fungi of the *Trichoderma* genus are considered one of the most used and promising in biological control [9]. This can be explained due to the versatility of the mechanisms of action against agricultural diseases and pests. *Trichoderma* species are capable of using different mechanisms of action, such as parasitism, production of metabolites (antifungal substances and antibiotics), production of polymer and protein degrading enzymes (glucanases, chitinases and proteases) [11]. In addition, these fungi can stimulate plant growth (production of phytohormones) and induce systemic resistance against diseases, which makes their use in agricultural systems even more promising [11].

Besides that, some isolates also have survival strategies that make them highly competitive in the environment, such as: survival in acid and / or saline soils; survival in conditions of high temperature and low humidity; fungicide resistance; adaptation to different environments and climatic zones, as they can inhabit soils in tropical regions and temperate climates; production of resistance structures; high efficiency in the use of resources as nutrients, thus making them excellent competing organisms; extraordinary capacity for proliferation in the rhizosphere and communication with plants, among others.

Due to the variation of environmental conditions (soil, climate, vegetation, etc.) on the planet, a species may have strains (variants) with specific adaptations to different environments. Thus, it is suggested that the antagonistic activity of a *Trichoderma* strain may change in relation to the same organism, if the fungus is used in regions with a microclimate different from the original fungus isolation site. This makes the search for antagonist agents from different regions relevant for obtaining isolates with potential application in biological control [12]. In this context, the objective of this work was to characterize and test fungi isolates of the *Trichoderma* genus to control *M. javanica*.

2. Material and methods

2.1 *Meloidogyne javanica* inoculum

The root-knot nematode inoculum, specifically *Meloidogyne javanica* (Est. J3), was obtained from a commercial soybean crop (*Glycine max*) in the municipality of Júlio de Castilhos, Rio Grande do Sul (29°04′55.5"S 53°41′07.7"W). Subsequently, the nematodes were extracted from the soybean roots by the method that consists of grinding in a blender with the addition of 0.5% sodium hypochlorite followed by sifting and centrifugation with sucrose solution, carried out at the Soil Biology laboratory of the Federal University of Santa Maria (UFSM). The extracted nematodes were inoculated in tomato plants cultivar "Santa Cruz" (*Solanum licopersycum* L.), to maintain the population. The tomatoes remained in a greenhouse with temperature controlled at 25°C ± 2°C. The females of the population were periodically submitted to electrophoresis with the esterase enzyme [13] to confirm the purity of the population.

2.2 In vitro evaluation of the nematicidal and nematostatic effect of *Trichoderma* spp. on *Meloidogyne javanica*

2.2.1 *Obtaining* Trichoderma *isolates*

Forty *Trichoderma* isolates from different regions of Brazil were used in this study showing the city that the isolates were provided (**Figure 1**). Isolates provided by private companies are from the Southeast and Midwest regions, however their locations were not provided.

Of the 40 isolates used in this study, 6 isolates are from the fungi bank of the laboratory of Soil Biology, 10 isolates were provided by the Federal University of Santa Maria, Frederico Westphalen (FW) campus, 6 isolates were supplied by the UFSM Phytosanitary Defense laboratory (D, DFS), 4 isolates were supplied by the University of Pelotas (Pel), 2 isolates were supplied by the UFSM campus Palmeira das Missões (PM), 1 isolate was supplied by the University of Passo Fundo (PF), 6 isolates were provided by Biota Innovations in the Midwest (BIF), and 5 control isolates were obtained from commercial products already reported as biological nematicides (**Table 1**).

The isolates were kept in Petri dishes containing the Potato-Dextrose-Agar (PDA) medium and incubated at 25°C ± 2°C in BOD (Biochemical Oxygen

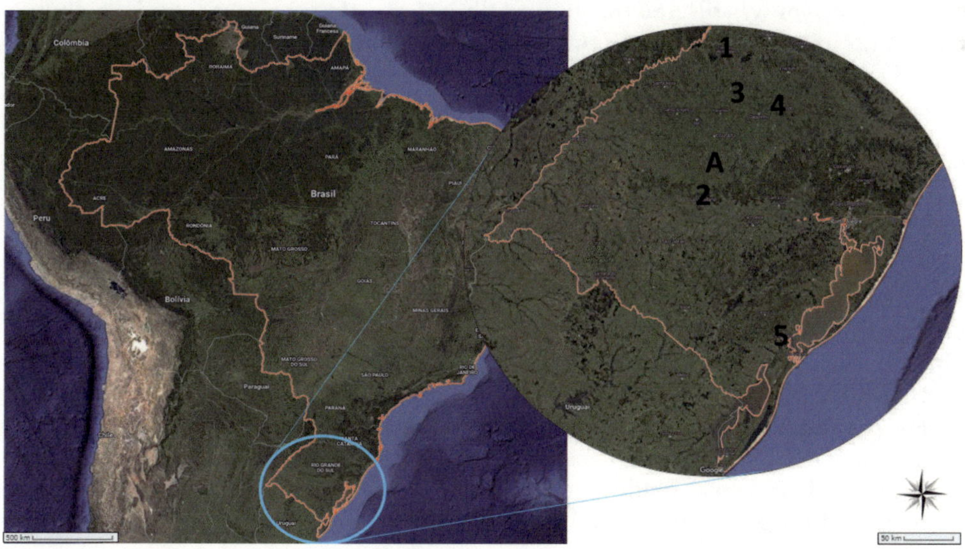

Figure 1.
Map of Brazil showing the provided locations of the isolates used in this work. The 1–5 are the cities-state where we obtained Trichoderma *isolates. The "A" is the location where the nematodes where collected. 1 – Frederico Westphalem-Rio Grande do Sul (RS); 2 – Santa Maria-RS; 3 – Palmeira das Missões-RS; 4 – Passo Fundo-RS; 5 – Pelotas-RS; A – Júlio de Castilhos-RS.*

Demand). Each specimen of the fungi was stored in inclined test tubes with PDA medium and kept refrigerated at a temperature of 6 to 10°C ± 2°C.

2.2.2 Parasitism test on M. javaniva eggs

For the parasitism test of the *Trichoderma* isolates on *M. javanica*, eggs of the nematode were extracted manually. These eggs were disinfected so that there was the least possible interference from other microorganisms that could be located on the outside of the eggs. For this, the eggs were placed in a test tube with 0.5% sodium hypochlorite solution, and stirred manually for one minute. After this stage, the eggs were also disinfected with 1% streptomycin and 0.1% 2-mercapto-ethanol (Sigma Aldrich) for four minutes; washed in sterile water and collected with micropipette.

From the obtained suspension, 50 eggs were added, and transferred to individual wells of ELISA plates. In each well, together with the J2, 100 μL of fungal suspension (10^8 conidia / ml) was added. Then the plates were kept in the dark in a BOD under a temperature of 25°C ± 2°C. Evaluations were performed 15 days after application. The numbers of parasitized eggs were determined. This test was repeated twice for greater data reliability.

2.2.3 Obtaining Trichoderma *filtrates*

Each fungus was grown in Petri dishes with PDA culture medium. Seven days after incubation at 25°C ± 2°C, three disks of 5 mm in diameter were removed from the edges of the cultures and placed in a 250 mL Erlenmeyer flask containing 100 mL of Czapek Dox liquid medium (0.5 g KCl, 1 g of KH_2PO_4, 2 g of $NaNO_3$, 30 g of sucrose, 0.01 g of $FeSO_4.H_2O$ and 0.5 g of $MgSO_4.7\ H_2O$ per 1000 mL of distilled water).

Erlenmeyer flasks were sterilized using the autoclave for 30 minutes. A different isolate was placed in each sterile Erlenmeyer. The flasks were kept in an incubator at 25°C with constant agitation for 15 days. After this period, the entire content of each Erlenmeyer was filtered through a cellulose acetate membrane, with an opening of 0.22 μm. For each isolate, the cellulose acetate membrane was exchanged. The fungal filtrates obtained were kept refrigerated for 48 hours at a temperature of 6 to 10°C ± 2°C, until the assay was established.

2.2.4 Nematode mortality and hatch inhibition test

For the mortality test of second stage juveniles (J2) of *M. javanica*, the Baermann funnel methodology modified by [14] was followed. The suspension with J2 was obtained from the hatching chamber with a tissue. From this suspension, 50 nematodes were removed through individual capture. In Elisa plates, 20 μL of water were pipetted and added to 80 μL of the fungal filtrates along with 50 captured nematodes. Mortality was assessed 48 hours after the application of the nematode suspension.

For the hatching test, first the egg suspension was obtained according to the methodology of [7]. Then we placed 50 eggs per well of the ELISA plate for the trial. The evaluation was made on the 21st day, when the count of 50 eggs was performed. In each treatment, eight repetitions were performed, kept at 25°C in the dark. These tests were repeated twice, aiming at increasing the data reliability. In the study, two controls were used, one containing only distilled water and the other containing only Czapek Dox medium, to eliminate the possibility of some type of unexplained alteration caused by the Czapek Dox medium that was used to perform the filtrates.

Isolates	Origin (Brazil)*	Treatment code	Species
FW09	South-FW	T1	T. asperellum
FW13	South-FW	T2	T. asperellum
FW14	South-FW	T3	*T. virens*
FW16	South-FW	T4	T. asperellum
FW21	South-FW	T5	T. asperellum
FW23	South-FW	T6	T. asperellum
FW31	South-FW	T7	*T. virens*
FW33	South-FW	T8	T. asperellum
FW36	South-FW	T9	*T. virens*
FW40	South-FW	T10	*T. virens*
UFSMQ1	South-SM	Til	*T. virens*
PM50	South-PM	T12	T. harzianum
PM63	South-PM	T13	T. harzianum
UFSM14	South-SM	T14	T. harzianum
PF102	South-PF	T15	T. harzianum
D33	South-SM	T16	T. asperellum
DFS03	South-SM	T17	*T. virens*
DFS04	South-SM	T18	T. asperellum
DFS05	South-SM	T19	T. asperellum
DFS06	South-SM	T20	T. harzianum
DFS07	South-SM	T21	T. asperellum
Pel210	South-Pel	T22	T. asperellum
Pel219	South-Pel	T23	T. harzianum
Pel221	South-Pel	T24	T. asperellum
Pel233	South-Pel	T25	T. harzianum
UFSMQ36	South-SM	T26	T. asperellum
UFSM27	South-SM	T27	T. asperellum
BIF0113	Southeast-**	T28	T. asperellum
BIF0111	Southeast-**	T29	T. asperellum
BIF0107	Southeast-**	T30	T. harzianum
BIF0119	Southeast-**	T31	T. asperellum
BIF0162	Southeast-**	T32	T. brevlcompactum
BIF0115	Southeast-**	T33	T. atroviride
UFSM34	South-SM	T34	T. asperellum
UFSM35	South-SM	T35	T. harzianum
CCT-7589	Midwest-**	T36	T. harzianum
SF-04	Midwest-**	T37	T. asperellum
12616	Midwest-**	T38	T. asperellum
T-22	Midwest-**	T39	T. harzianum
ESALQ-1306	Southeast-**	T40	T. harzianum

**FW: Frederico Westphalem-RS; SM: Santa Maria-RS; PM: Palmeira das Missões-RS; PF: Passo Fundo-RS; Pel: Pelotas-RS.*
***The city of origin of these isolates are not known or were not disclosed by the provider.*

Table 1.
Trichoderma *isolates obtained from different regions of Brazil.*

2.2.5 Experimental design and statistical analysis

The experimental design used was completely randomized with eight replicates for each treatment, and each fungal isolate corresponded to one treatment (40 isolates). The variables evaluated were: number of live and dead J2 nematodes, count of J2 hatching, count of the number of parasitized and darkened eggs. The results were subjected to analysis of variance, and the means of each treatment were compared by the Scott-Knott cluster test at 5% probability of error, by the SISVAR software [15].

2.2.6 Molecular identification of Trichoderma *spp.*

The total genomic DNA was extracted by the method described by [16]. The *Trichoderma* isolates were placed in a 1.5 mL microtube with 400 μL of extraction solution (Tris–HCl 100 mM pH 8.0; EDTA 20 mM pH 8.0; NaCl 1.4 M; CTAB 2%; PVP 1%; 2-Mercaptoethanol 0.1% and Proteinase K 0.01%) previously heated to 65°C for 3 min. and vortexed for 10 seconds. Then, it was incubated in a water bath at 65°C for 45 minutes, shaking every 15 min.

Then 400 μL of chloroform was added and stirred by gentle inversions for 5 min. Afterwards, it was centrifuged at 14000 rpm, 20°C, for 5 min. After centrifugation, approximately 400 μL of the aqueous phase was removed and transferred to a new 1.5 mL microtube, where 200 μL of chilled isopropanol (2-propanol) was added and homogenized by gentle inversions for 1 minute and incubated at −20°C for 30 min. The solution was centrifuged at 1400 rpm, 20°C, for 5 min. The supernatant was discarded, keeping only the pellet at the bottom of the microtube. For DNA precipitation, 200 μl of cold 70% ethanol (4°C) was added to the tube, followed by centrifugation at 14000 rpm at 4°C, for 5 min. and the supernatant was discarded keeping the pellet formed. The precipitate was dried at room temperature, and recovered in a volume of 50 μL TE [1 mM Tris and 0.1 mM EDTA] + RNAse and incubated at 37°C for 30 min., and its DNA was quantified and stored at −20°C until use.

The genomic DNA samples extracted from the fungi were subjected to polymerase chain reaction (PCR) with that performed for partial amplification of the elongation factor gene (EF-1α) with the primers 5'-ATGGGTAAGGARGACAAGAC-3′ and 5'-GGARGTACCAGTSATCATGTT3-'. For this, 3 μL of the fungi DNA were added to the final volume of the 25 μL PCR reaction, containing 10 mM Tris HCl pH 8.3; 50 mM KCl; 1.1 mM $MgCl_2$; 10 mM of each dNTP; 25 nmoles of each EF1 and EF2 primer; 1.5 μL of Taq DNA polymerase (Invitrogen, Brazil) and ultrapure water to complete the reaction volume. A negative control without DNA was included in the PCR. The amplification reactions were carried out in a thermocycler (Applied Biosystems 2720, Thermo Fisher Scientific, USA), under the following conditions: 94°C for 1 min., 35 cycles of 95°C for 3 min., 95°C for 1 min., 72°C for 1 min. and 30 seconds, and 72°C for 10 min. At the end of the reaction, the amplified fragments were kept at 4°C. To verify amplification, electrophoresis was performed on 1.5% agarose gel, in TBE 1X buffer, stained with Sybr Gold (Invitrogen, Brazil). PCR products were purified with the Gen Elute PCR clean-up Kit® kit (Sigma, USA) and sequenced (ABI PRISM 3100, Thermo Fisher Scientific, USA). The sequences were analyzed using the Staden Package 2.0.0b program [17] to obtain consensus.

2.2.7 Phylogenetic analysis

The alignment of the nucleotide sequences was performed in the programs Clustal W and Clustal X [18], and sequences deposited in the databases were used for comparisons. The Neighbor-joining method, using the Jukes-Cantor model, was used to estimate the evolutionary distance. The phylogenetic tree was built in the MEGA X program [19], with the Maximum Likelihood algorithm and the bootstrap values calculated with 1,000 replicates.

3. Results and discussion

3.1 Molecular identification of *Trichoderma* spp.

The *Trichoderma* isolates used in the study were identified from DNA extraction and separated into five distinct species, with 20 isolates belonging to the *T. asperellum* (50% of the isolates) species, 12 isolates of *T. harzianum* (30%), six isolates of *T. virens* (15%), an isolate of *T. brevicompactum* (2.5%) and an isolate belonging to the *T. atroviride* (2.5%) species.

The isolates T36 to T40 are commercially used *Trichoderma* spp. The T36, T39 and T40 were identified as *T. harzianum*, and T37 and T38 as *T. asperellum*. Using these isolates as references we compared with the sequences of other isolates and the clusters that resulted are in the phylogenetic tree (**Figure 2**).

Analyzing the bootstrap values presented in the dendrogram, we can say that within the clades of the species *T. asperellum*, *T. virens* and *T. harzianum*, we obtained sequences of small size, which caused low bootstrap values, as there was no enough support for the node. In [20], the authors reported that bootstrap values below 70 are normally hidden, and above 70 usually correspond to probabilities greater than 95% that the clade is real.

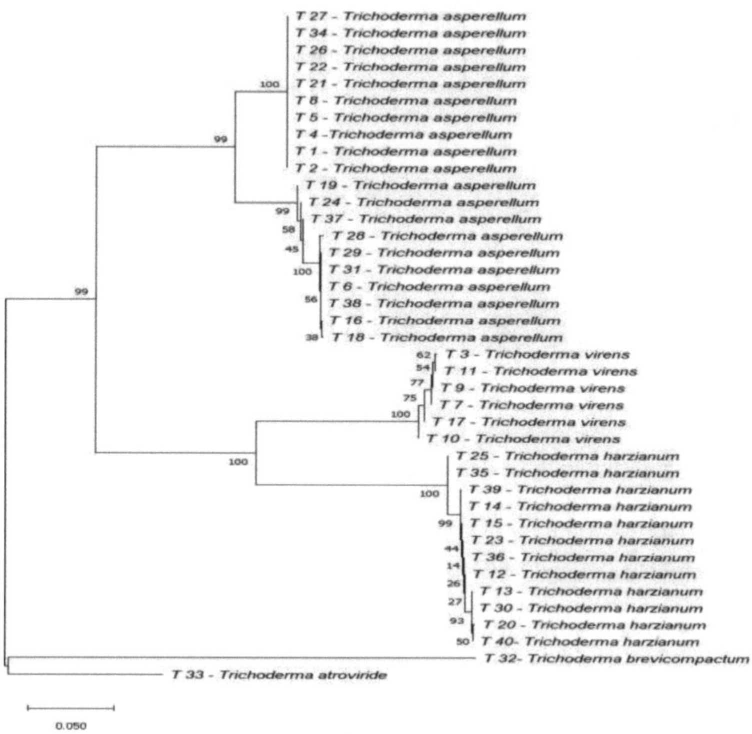

Figure 2.
Dendogram of the partial sequences of the EF-1α elongation factor (TEF) gene of Trichoderma *spp. using the maximum likelihood method based on the Jukes-Cantor model.*

3.2 In vitro evaluation of the parasitic, nematicidal and nematostatic effect of *Trichoderma* spp. on *M. javanica*

In this study, it was observed that all species of *Trichoderma* tested showed high potential for suppression of *M. javanica* (**Table 2**). This was confirmed by the effect of parasitism of eggs and nematodes and also by the production of metabolites that killed and/or inhibited the hatching of J2 eggs. The percentage of parasitized J2 nematodes was greater than 85% for all tested isolates, and 35% (14) of the isolates showed a parasitism capacity greater than 95%. The isolates that stood out with the highest parasitism averages, forming a group statistically different from the other isolates, were: *T. harzianum* ESALQ-130 (99%), *T. virens* DFS03 (98.8%), *T. harzianum* UFSM14 (98.7%), *T. atroviride* BIF0115 (98.7%), *T. asperellum* 12616 (98.4%), *T. asperellum* SF-04 (98.4%), *T. harzianum* T-22 (98.3%) %), *T. asperellum* UFSM27 (98.4%), *T. asperellum* BIF0119 (97.9%) and *T. harzianum* PF102 (96.8%).

Regarding the mortality of *M. javanica* using fungal filtrates (**Table 2**), it was observed that 16 (40%) of the isolates had a mortality rate ranging from 85% to 93.1%. Thus, the *Trichoderma* isolates that had the highest mortality values were: *T. harzianum* CCT-7589 (93.1%), *T. brevicompactum* BIF0162 (92.2%), *T. asperellum* BIF0111 (91.9%), *T. atroviride* BIF0115 (91.9%), *T. asperellum* UFSM27 (91.8%), *T. asperellum* BIF0111 (91.7%), *T. harzianum* BIF0107 (91.7%), *T. harzianum* ESALQ-1306 (91.7%), *T. asperellum* UFSM34 (91.6%), *T. harzianum* UFSM35 (91.0%), *T. asperellum* SF-04 (90.9%), *T. asperellum* 12616 (90.8%), *T. harzianum* T-22 (90.6%), *T. harzianum* PM63 (90.1%), *T. virens* DFS03 (88.7%), *T. asperellum* Pel221 (85.8%) and *T. asperellum* BIF0119 (85%). In contrast, the isolates that obtained the lowest percentages of J2 mortality from *M. javanica* were *T. asperellum* FW16 and *T. asperellum* DFS04, with 70.8 and 65.3% mortality, respectively.

Results similar to the present study were obtained by [21], where all the filtrates obtained from the isolates of *Trichoderma* spp. (22 isolates) proved to be efficient in promoting juvenile mortality in a population of *M. incognita* after 24 hours of application of the filtrates in vitro. It is reported that the parasitic effect of fungi of the genus Trichoderma against nematodes from egg to adult stages, requires some facilitating mechanism [22], such as the presence of lytic enzymes. The mortality of J2 de Meloidogyne may be related to the presence of enzymes proteases and chitinases that act in the degradation of the cuticle of nematodes, a resistant coating structure composed of proteins and chitin. For this reason, metabolites obtained from fungal isolates that may have these facilitating mechanisms are also sought.

Trichoderma isolates produce metabolites, such as lytic enzymes, which are released into the rhizosphere solute, in our case, culture medium solute. Thus, the inhibition of J2 hatching from the application of *Trichoderma* filtrates may be related to enzymatic activity in the solution. Meloidogyne eggs are composed of approximately 30% chitin in the outer layer, in addition to other structural proteins, and lipids in the inner layer. In this sense, the action of inhibiting the hatching of J2 indicates the presence of enzymes proteases, chitinases and lipases that acted on the enzymatic degradation of the outer and inner layers of the eggs of *M. javanica*, reducing the hatching capacity of J2 of the eggs.

Regarding the nematicidal and nematostatic effect in the inhibition of the hatching of J2 by fungal filtrates, it was observed that 25% of the isolates resulted in a high inhibition of the hatching of *M. javanica* J2, varying from 91.6% to 94.4% hatching inhibition (**Table 2**). The isolates that showed the highest percentage of inhibition statistically were: *T. harzianum* CCT-7589 (94.4%), *T. asperellum* UFSM34 (94.0%), *T. asperellum* SF-04 (93.8%), *T. harzianum* BIF0107 (93.4%), *T. atroviride* BIF0115 (92.6%), *T. brevicompactum* BIF0162 (92.5%), *T. harzianum* T-22 (91.9%), *T. harzianum* UFSM35 (91.8%), *T. harzianum* ESALQ-1306 (91.8%) and

Treatments	Species	J2 parasitized (%)		J2 mortality (%)		J2 hatching inhibition (%)	
T1	Trichoderma asperellum	92.62	C	74.9	C	84.6	C
T2	Trichoderma asperellum	90.87	C	80.4	B	81.5	C
T3	Trichoderma virens	87.81	E	81.7	B	73.1	D
T4	Trichoderma asperellum	91.56	C	70.8	D	84.9	C
T5	Trichoderma asperellum	93.37	C	81.5	B	79.4	C
T6	Trichoderma asperellum	88.12	E	84.3	B	80.6	C
T7	Trichoderma virens	91.25	C	82.5	B	83.5	C
T8	Trichoderma asperellum	87.62	E	77.9	B	86	C
T9	Trichoderma virens	85.5	E	75.9	C	83.6	C
T10	Trichoderma virens	93.37	C	80.3	B	66	E
T11	Trichoderma virens	94.62	B	79.2	B	84.8	C
T12	Trichoderma harzianum	90.43	C	79.4	B	85.3	C
T13	Trichoderma harzianum	89.87	D	90.1	A	87.3	B
T14	Trichoderma harzianum	98.68	A	83	B	87.2	B
T15	Trichoderma harzianum	96.81	A	84.2	B	86.7	B
T16	Trichoderma asperellum	92.5	C	75.5	C	77.7	C
T17	Trichoderma virens	98.81	A	88.7	A	87.2	B
T18	Trichoderma asperellum	95	B	65.3	D	85.3	C
T19	Trichoderma asperellum	93.31	C	74.8	C	85.5	C
T20	Trichoderma harzianum	94.81	B	79.3	B	83.6	C
T21	Trichoderma asperellum	94.18	B	78.9	B	81.5	C
T22	Trichoderma asperellum	95.75	B	80.2	B	79.9	C
T23	Trichoderma harzianum	92.62	C	80.8	B	79.7	C
T24	Trichoderma asperellum	94.06	B	85.8	A	88.7	B
T25	Trichoderma harzianum	94.87	B	81.2	B	85.2	C
T26	Trichoderma asperellum	92.18	C	82.4	B	83.8	C
T27	Trichoderma asperellum	98.43	A	91.8	A	88.3	B
T28	Trichoderma asperellum	89.75	D	74.8	C	82.9	C
T29	Trichoderma asperellum	89.68	D	91.9	A	88.5	B
T30	Trichoderma harzianum	93.93	B	91.7	A	93.4	A
T31	Trichoderma asperellum	97.93	A	85	A	87.2	B
T32	Trichoderma brevicompactum	92.62	C	92.2	A	92.5	A
T33	Trichoderma atroviride	98.68	A	91.9	A	92.6	A
T34	Trichoderma asperellum	96.18	B	91.6	A	94	A
T35	Trichoderma harzianum	94.12	B	91	A	91.8	A
T36	Trichoderma harzianum	95.68	B	93.1	A	94.4	A
T37	Trichoderma asperellum	98.43	A	90.9	A	93.8	A
T38	Trichoderma asperellum	98.43	A	90.8	A	91.6	A
T39	Trichoderma harzianum	98.31	A	90.6	A	91.9	A

Treatments	**Species**	**J2 parasitized (%)**		**J2 mortality (%)**		**J2 hatching inhibition (%)**	
T40	Trichoderma harzianum	99	A	91.7	A	91.8	A
Control H_2O*		0	F	6.3	E	8	F
Control Czapek Dox**		—		5	E	7.1	F
CV (%)		4.37		6.9		7.8	

Means followed by different capital letters are differentiated by the Scott-Knott hierarchical cluster test with 95% confidence.
**Only used H_2O as control treatment.*
***Only used Czapek Dox medium as control treatment.*

Table 2.
Effect of parasitism of Trichoderma *spp. and nematicidal and nematostatic effect of metabolites of fungal filtrates in* Meloidogyne javanica.

T. asperellum 12616 (91.6%). In contrast, the isolate *T. virens* FW40 resulted in the lowest inhibition capacity of J2 hatching (66% inhibition).

In [23], they concluded that two *Trichoderma* isolates (*T. asperellum* M2RT4 and *Trichoderma* sp. MK4) significantly reduced the number of hatched J2s, between 60.8 and 81.8%. In the same work, it was observed that *T. asperellum* M2RT4 was the most efficient isolate for the control of galls, egg mass and deposited eggs, reducing, respectively, 81.8, 78.5 and 88.4%, indicating that isolates from this species have potential for biocontrol. Both results coincide with data on egg and J2 mortality and inhibition of hatching of J2, obtained in this work. On the other hand, *T. atroviride* F5S21 had no significant effect in comparison to the control, which goes against the result obtained in the present study, whereas *T. atroviride* (T33) showed positive results in parasitism, mortality and inhibition of hatching of J2.

The authors [24] mention that *Trichoderma* filtrates have toxic effect on adults of *Meloidogyne* sp. The same was observed by [25], in which all filtrates showed toxic activity against *M. incognita*, obtaining 98% of immobile and dead J2.

In a study by [26] with the objective of evaluating the effect of *T. harzianum* for the biological control potential of *M. javanica*, they also found promising results for nematode suppression. In this study, they also evaluated the metabolite production capacity and its relationship to nematode suppression. The authors concluded that the increased activity of the chitinase enzyme is directly related to the capacity of suppression of *M. javanica* by the species *T. harzianum*.

The production of lytic enzymes also helps in the penetration of the fungus, especially visualized in *T. harzianum* [27, 28]. The ability to produce metabolites, as we can see, is also an important factor in the suppression of nematodes [29]. In our study, the production of metabolites by fungi, through fungal filtrates, also had a positive result for biocontrol of *M. javanica*. The production of metabolites by the fungi inhibited the outbreak of J2, and the isolate that obtained the lowest percentage of inhibition of J2 was a *T. virens* isolate. However, we can observe that this same species had percentages above 80% of mortality, parasitism and inhibition of hatching, which shows us an important difference between the *Trichoderma* isolates.

Results from [30] showed that, fungi of the *Trichoderma* genus, specifically *T. harzianum* in this study, presented positive results for the mortality and hatch inhibition of J2 comparing with control, *Paecilomyces lilacinus* and humic acid. And [31], also observed for 3 different isolates of *Trichoderma* 100% larval mortality for *M.*

javanica and between 94,1 and 100% larval mortality of *M. incognita*. Both studies support our findings for the potential of *Trichoderma* for *Meloidogyne* mortality.

As we can see in the results of this work and supported by others, there is a significant difference in the suppression of nematodes among isolates of the same species. We observed in our work species of *T. asperellum* that ranged from 87.62% of parasitized J2, up to 98.43%, and also, other strains of the same species showed 65.3% of J2 mortality, while others showed 91.9%, and for the percentage of inhibition of J2 hatching, there was a variation from 77.7–94%, within the same species. In the species *T. harzianum*, there were variations of 89.87–99% in parasitism, of 79.3% to 93.1% of J2 mortality from the filtrates, and from 79.7% to 94.4% of hatching inhibition, in different strains.

This is due to the fact that each strain has a different gene expression, that is, it has genetically the ability to produce different lytic enzymes [22, 32], or it has the parasitic capacity expressed by virulence genes. For this reason, selections of highly efficient strains are carried out in the biocontrol of pests and diseases.

It was tested by [12] 230 isolates, and only 8 belonging to the South region. All organisms obtained in prospecting processes should be tested as potential antagonistic agents for different pathogens and pests. The authors affirm the importance of prospecting for biological control agents and characterizing them, mainly fungi, in the entire Brazilian territorial area, since good antagonists may be dispersed in different regions of the country. This work shows the potential of five species of *Trichoderma* in the biological control of *M. javanica*, and future experiments in natural conditions could help to widen the differences of each of the studied strains.

4. Conclusion

All *Trichoderma* isolates tested showed potential for *Meloidogyne javanica* biocontrol.

Acknowledgements

We thank the Federal University of Santa Maria and the Graduate Program in Soil Science, for offering the entire structure for the development of the work. To CNPQ (National Council for Scientific and Technological Development) for the 12 months of scholarship.

Conflict of interest

The authors hereby declare no conflict of interest.

Author details

Ricardo R. Balardin[1*], Cristiano Bellé[2], Daiane Dalla Nora[1], Rodrigo F. Ramos[1], José Carlos V. Rodrigues[3] and Zaida I. Antoniolli[1]

1 Federal University of Santa Maria, Santa Maria, Brazil

2 Phytus Institute, Santa Maria, Brazil

3 University of Puerto Rico, San Juan, Puerto Rico

*Address all correspondence to: ricardorbalardin@gmail.com

References

[1] Dias WP, Garcia A, Silva JFV, Cameiro GEDS. Nematóides em soja: identificação e controle. Embrapa Soja. Circ. 76 [Internet]. 2010;1-8. Available from: http://agris.fao.org/agris-search/search/display.do?f=2012/BR/BR2012113100131.xml;BR20101854178

[2] Almeida AMR, Ferreira LP, Yorinori JT, Silva JFV, Henning AA, Godoy CV, Costamilan LM & Meyer MC. Doenças da soja. In: Kimati H, Amorim L & Rezende JAM (Eds.) Manual de fitopatologia: doenças das plantas cultivadas. São Paulo, Agronômica Ceres. 2005;4:570-588.

[3] Mattos V. Variabilidade genética e agressividade a soja de populações de *Meloidogyne* spp. do Cerrado e de áreas de cultivo. 2013; Available from: http://repositorio.unb.br/handle/10482/14040

[4] Kirsch VG, Kulczynski SM, Gomes CB, Calderan A, Gabriel M, Bellé C, et al. *Helicotylenchus* Associadas À Soja No Rio Grande Do Sul. Nematropica. 2016;46:197-208.

[5] Moens, M, Roland NP, and James LS. *Meloidogyne* species–a diverse group of novel and important plant parasites. Root-knot nematodes 2009;1:483.

[6] Coyne DL, Cortada L, Dalzell JJ, Claudius-Cole AO, Haukeland S, Luambano N, et al. Plant-parasitic nematodes and food security in Sub-Saharan Africa. Annu. Rev. Phytopathol. 2018;56:381-403.

[7] Hussain MA, Mukhtar T, & Kayani MZ. Efficacy evaluation of *Azadirachta indica*, *Calotropis procera*, *Datura stramonium* and *Tagetes erecta* against root-knot nematodes Meloidogyne incognita. Pak. J. Bot, 2011;43(1):197-204.

[8] Singh V, Mawar R, & Lodha S. Combined effects of biocontrol agents and soil amendments on soil microbial populations, plant growth and incidence of charcoal rot of cowpea and wilt of cumin. Phytopathologia Mediterranea, 2012:307-316.

[9] Sikora RA, Schäfer K, & Dababat AA. Modes of action associated with microbially induced in planta suppression of plant-parasitic nematodes. Australasian Plant Pathology, 2007;36(2):124-134.

[10] de Freitas Soares FE, Sufiate BL, & de Queiroz JH. Nematophagous fungi: Far beyond the endoparasite, predator and ovicidal groups. Agriculture and Natural Resources, 2018;52(1):1-8.

[11] Monte E, Bettiol W, & Hermosa R. *Trichoderma* e seus mecanismos de ação para o controle de doenças de plantas. Trichoderma, 2019:181.

[12] Louzada GADS, Carvalho DDC, Mello SCM, Lobo Júnior M, Martins I, & Braúna LM. Potencial antagônico de *Trichoderma* spp. originários de diferentes agroecossistemas contra *Sclerotinia sclerotiorum* e *Fusarium solani*. Biota neotropica, 2009;9(3):145-149.

[13] Carneiro RM, Moreira WA, Almeida MR, & Gomes ACM. Primeiro registro de Meloidogyne mayaguensis em goiabeira no Brasil. Nematologia Brasileira, Brasília, 2001;25:223-228.

[14] Christie JR, & Perry VG. Removing nematodes from soil. Proceedings of the Helminthological Society of Washington, 1951;18(2):106-108.

[15] Ferreira DF. Sisvar: a computer statistical analysis system. Ciência e agrotecnologia, 2011;35(6):1039-1042.

[16] Doyle JJ, & Doyle JL. Isolation ofplant DNA from fresh tissue. Focus, 1990;12(13):39-40.

[17] Staden R, Judge DP, & Bonfield JK. Analyzing sequences using the Staden Package and EMBOSS. In Introduction to bioinformatics. Humana Press, Totowa, NJ, 2003:393-410.

[18] Thompson JD, Gibson TJ, Plewniak F, Jeanmougin F, & Higgins DG. The CLUSTAL_X windows interface: flexible strategies for multiple sequence alignment aided by quality analysis tools. Nucleic acids research, 1997;25(24):4876-4882.

[19] Kumar S, Stecher G, Li M, Knyaz C, & Tamura K. MEGA X: molecular evolutionary genetics analysis across computing platforms. Molecular biology and evolution, 2018;35(6):1547.

[20] Hillis DM, & Bull JJ. An empirical test of bootstrapping as a method for assessing confidence in phylogenetic analysis. Systematic biology, 1993;42(2):182-192.

[21] Freitas MA, Pedrosa EMR, Mariano RLR, & Maranhão SRVL. Seleção de *Trichoderma* spp. como potenciais agentes para biocontrole de *Meloidogyne incognita* em cana-de-açucar [Screening *Trichoderma* spp. as potential agents for biocontrol of *Meloidogyne incognita* in sugarcane]. Nematropica, 2012:115-122.

[22] Szabó M. Potential of *Trichoderma* species and nematodetrapping fungi to control plant–parasitic nematodes: in vitro confrontation and gene expression assays using *Caenorhabditis elegans* model system (Doctoral dissertation, Ph. D. thesis. Szent Istv án University, Gödöllo, Hungary), 2014.

[23] Kiriga AW, Haukeland S, Kariuki GM, Coyne DL, & Beek NV. Effect of *Trichoderma* spp. and *Purpureocillium lilacinum* on *Meloidogyne javanica* in commercial pineapple production in Kenya. Biological Control, 2018;119:27-32.

[24] Bokhari FM. Efficacy of some *Trichoderma* species in the control of *Rotylenchulus reniformis* and *Meloidogyne javanica*. Archives of Phytopathology and Plant Protection, 2009;42(4):361-369.

[25] Costa MJ, Campos VP, Pfenning LH, & Oliveira DF. Toxicidade de filtrados fúngicos a *Meloidogyne incognita*. Fitopatologia Brasileira, 2001;26:749-755.

[26] Sahebani N, & Hadavi N. Biological control of the root-knot nematode *Meloidogyne javanica* by *Trichoderma harzianum*. Soil biology and biochemistry, 2008;40(8):2016-2020.

[27] Spiegel Y, & Chet I. Evaluation of Trichoderma spp. as a biocontrol agent against soilborne fungi and plant-parasitic nematodes in Israel. Integrated Pest Management Reviews, 1998;3(3):169-175.

[28] Albehadeli Y, Mamarabadi M, & MahdiKhani E. Possibility of the biocontrol of *Meloidogyne javanica* using the fungus *Trichoderma harzianum* under greenhouse condition. Plant Archives, 2019;19.

[29] Goswami J, Pandey RK, Tewari JP, & Goswami BK. Management of root knot nematode on tomato through application of fungal antagonists, *Acremonium strictum* and *Trichoderma harzianum*. Journal of Environmental Science and Health Part B, 2008;43(3):237-240.

[30] Al-Hazmia AS, Al-Yahyaa FA, AbdelRafaab OA, & Lafia AH. Effects of humic acid, Trichoderma harzianum and Paecilomyces lilacinus on Meloidogyne javanica. Int J Agric Environ Biores, 2019;4(1):61-74.

[31] Migunova V, Sasanelli N, & Kurakov A. Effect of microscopic fungi on larval mortality of the root-knot nematodes Meloidogyne incognita and

Meloidogyne javanica. Biological and integrated control of plant pathogens IOBC-WPRS Bulletin, 2018;133:27-31.

[32] Szabó M, Csepregi K, Gálber M, Virányi F, & Fekete C. Control plant-parasitic nematodes with Trichoderma species and nematode-trapping fungi: The role of chi18-5 and chi18-12 genes in nematode egg-parasitism. Biological Control, 2012;63(2):121-128.

Chapter 9

Biological Control of Root-Knot Nematodes Using *Trichoderma* Spp.

Linnley Mulusa

Abstract

Agriculture is an important activity globally since it ensures food security and is a source of income for many families, especially those living in underprivileged countries. The continuous growth in the global population has seen farmers increase the crop production acreage to meet the increasing demand for food and avert food shortage. Despite this, farmers continue to harvest lower yields than anticipated, which threatens global food security. The reduced yields result from outdated and ineffective farming practices as well as pests and diseases. Diseases are a significant cause of reduced crop yields globally. Biotic and abiotic factors cause diseases. Of the recognized biotic causes of disease, root-knot nematodes, also known as *Meloidogyne* spp. are plant-parasitic nematodes that cause significant losses to farmers in terms of reduced plant yields. Over the years, researchers have conducted several studies on the effective use of *Trichoderma* spp. fungi as a biocontrol agent for these pathogens. This paper analyzes the advancements made towards the effective and efficient biocontrol of *Meloidogyne* spp. using *Trichoderma* spp. and the implications of these advancements for agriculture and food security.

Keywords: Food security, root-knot nematodes, *Trichoderma* spp., biocontrol, plant growth promotion

1. Introduction

Agriculture is an important economic activity globally because almost 65% of working adults in underprivileged populations earn their livelihood from agriculture and it contributes significantly to the global gross domestic product (GDP) [1]. Since farmers engage in both cash crop and subsistence farming, they can either use their agricultural produce locally or export it. Due to the high demand for agricultural produce, farmers have increased the acreage of land used to grow their crops [2]. However, the increase in acreage of the land used for production is indirectly proportional to the yields obtained from the fields [2]. These results imply that the technologies used by the farmers either are becoming obsolete or applied inefficiently. Another implication is that the farmers do not use their land efficiently [1]. With the continuous increase in the global population comes the increased demand for food to sustain the population. Therefore, farmers need to employ better and more efficient farming technologies on their farms and use their land resourcefully. Failure to remedy the situation might result in a food crisis worldwide, and especially in populous developing countries.

Pests and diseases reduce crop yields on farms for several crops like potato, millet, maize, rice, wheat, and soybean. Controlling plant pests and diseases is difficult due to various factors including the co-evolution of pests, pathogens, and plants over the years in addition to the high numbers and large diversity of the pests and pathogens. Categories of pests that attack crops include rodents, insects, and birds [3]. Insects damage crops either directly or indirectly. Direct damage involves insect activity that causes injury to crops such as burrowing holes in different plant tissues and feeding on them [3]. The resulting damage to plant tissues disrupts the physiological activities of plants, for example, photosynthesis and water uptake and this leads to the decreased yields. Insects, mainly aphids, can also cause indirect damage to crops by acting as vectors of plant-parasitic microorganisms, for example, nematodes. Birds and rodents cause direct damage to crops by feeding on different crop tissues. The indiscriminate feeding habits of rodents is a significant challenge to farmers since the pests can feed on any crop grown in the field. If farmers are to improve their yields, they must identify the pests in their farms and use effective technologies for control.

Diseases, on the other hand, arise from biotic and abiotic factors. Crop diseases affect the physiological activity of plants resulting in a significant reduction in crop yields in addition to unprecedented costs for farmers. Abiotic factors that cause plant diseases include unfavorable growth conditions such as inadequate nutrients and insufficient sunlight and mesobiotic factors [4]. Mesobiotic factors are entities that exhibit an intermediate state of living and non-living organisms. These entities include viroids and viruses. Biotic factors that cause plant diseases are animate and pathogenic. These organisms are both eukaryotes and prokaryotes. Prokaryotic organisms are of bacterial origin and cause diseases such as soft rot in vegetables and wilting in potatoes. Eukaryotes that cause plant diseases include fungi (powdery mildew, rust and smut), protozoa (phloem necrosis in coffee and hart rot in palm and coconut trees), algae (red rust in mango and papaya trees), and nematodes (root-knot in vegetables, ear cockle of wheat, and molya disease in wheat and barley) [4]. Apart from reducing the crop yield, plant diseases can also result in a disruption of the natural ecosystem, causing an imbalance of living organisms in the environment.

Further, different types of *Meloidogyne* spp., a polyphagous, endoparasitic and sedentary root-knot nematode, attack plants in the field resulting in reduced yields. These include *M. chitwoodi*, *M. incognita* and *M. javanica*, also known as the southern root-knot nematode [5, 6]. These nematodes produce aboveground and belowground symptoms in the affected plants. These symptoms affect plant growth and development and therefore reduce yields [6]. It is therefore vital to control the nematode population of *Meloidogyne* spp. in farms to prevent the occurrence of a global food crisis. Due to the high economic impact of root-knot nematodes, companies, researchers, and farmers have developed several strategies for controlling them [7]. One strategy involves the use of chemical nematicides. However, since these chemicals are detrimental to human health and pollute the environment, authorities and other concerned organizations increasingly discourage their application [7, 8]. Another method used by farmers is crop rotation. The disadvantage of using this method is that nematodes might form dauer stages that enable them to survive in the soil until they detect a potential host and infect it again. Strategies related to biocontrol are safer and more effective for the management of root-knot nematodes because biocontrol agents target specific organisms and reduce their populations and this reduces the damage to crops and the costs farmers incur when purchasing broad-spectrum nematicides.

The use of biological control agents to mitigate the damage caused by root-knot nematode infestation has several advantages. These advantages are of significant value to food crop and cash crop farmers and the environment. First, biocontrol agents are specific to the target organism; therefore, they do not destroy other beneficial organisms in the process [3]. Biocontrol can also provide a long-term solution to crop pests, reducing the costs required for pest control on the farm. Additionally, the biocontrol agents do not cause environmental pollution; therefore, their application does not harm other organisms and humans in the environment [3]. Further, unlike chemicals, pests do not develop resistance to biological control agents. It is possible to control root-knot nematodes at different stages of their lifecycle; therefore, the farmer can apply the biocontrol agent depending on the nematode stage and the agent's effectiveness at that stage [7]. One of these stages is the hatching stage. Given that the eggs produce the infective stages of the endoparasite *M. incognita,* it would be more desirable to control the nematodes at this stage before they get to the plants and damage them. Another stage of plant-parasitic nematode control is the infective J2 larva stage. Controlling *M. incognita* larvae at their infective stage ensures that the nematodes do not infect susceptible host plants and establish feeding sites from where they absorb plant nutrients and develop into adults, which produce subsequent generations. Research also shows that biocontrol agents can control adult root-knot nematodes.

According to research, different microbes are viable options for the biocontrol of root-knot nematodes. These microorganisms use different mechanisms including the production of toxins and antibiotics, parasitism, and boosting a plant's immunity for biocontrol [8–10]. Since *Trichoderma* spp. fungi have a double effect – acting as biocontrol agents and promoting plant growth, this chapter focuses on the genus, comprising free-living microbes found in the soil. Several studies indicate that the fungi is a potential biocontrol agent against the root-knot nematodes in the *Meloidogyne* genus, which are a global menace. For the effective application of the fungi as a biocontrol agent, it is important to understand its biological and physiological processes because a biocontrol agent should compete favorably and persist in the environment. Additionally, the agent should colonize newly formed roots rapidly, multiply on the roots efficiently, and benefit the colonized crop continuously until harvest.

2. Root-knot nematodes

The dynamic relationship between nematodes and their plant hosts resulted in the evolution of plant-parasitic nematodes over time [11]. Due to these evolutions, the nematodes acquired favorable characteristics for their survival and development. These characteristics include the development of feeding structures like the stylet that differentiates plant-parasitic nematodes from other nematodes and secretions that are essential for infecting the plant host and absorbing nutrients [11]. Currently, there are more than 4,100 species of plant-parasitic nematodes [6, 8]. These microorganisms cause damage to crops and this has significant economic implications for farmers and consumers. On average, farmers around the world incur losses that range between 80 and 118 billion dollars each year because of the activity of plant-parasitic nematodes [6]. Fifteen percent of plant-parasitic nematodes that have a huge economic impact are those that infect the roots of the host plant and therefore hinder the uptake of water and nutrients by the plant [6].

The most successful plant-parasitic nematode species are sedentary endoparasitic nematodes. These nematodes establish a permanent feeding site within the roots of the host and obtain nutrients from this site while completing their lifecycle [6, 11]. The major genera of plant-parasitic nematodes associated with significant crop damage and losses are *Xiphinema, Pratylenchus, Hoplolaimus, Heterodera,*

Rotylenchulus, and Meloidogyne. Crops affected by these nematodes include wheat, finger millet, rice, maize, potato, and sweet potato [6]. Root-knot nematode species belonging to the *Meloidogyne* genera are the most dominant root-knot nematodes [11]. The genus has over 100 species. Of these species, the ones that cause the most devastating damage to crops include *M. javanica*, *M. halpa, M. graminicola, M. chitwodii, M. arenaria*, and *M. incognita*. *Meloidogyne* spp. have a global distribution especially in the tropical and subtropical regions and a wide host range [6].

Root-knot nematodes (RKNs), whose scientific name is *Meloidogyne* spp. are polyphagous microorganisms, meaning they have a wide host range. The pathogens belong to the order Tylechnida and family Meloidogynidae and inhabit the soil [11, 12]. Environmental factors that promote the development and presence of *Meloidogyne* spp. are warm temperatures and moist soils. The optimum temperature for root penetration by *Meloidogyne* spp. is 27°C; however, depending on the species, this temperature ranges from 10 to 25°C [12]. Female RKNs lay eggs when the temperature ranges from 14.2–31.7°C [12]. Under these temperatures, the females lay 300 to 800 eggs in a gelatinous matrix [12]. If the conditions are favorable, RKNs can produce a new generation within 25 days; therefore the pathogens can produce multiple generations within a year or during a planting season, which increases the intensity of the damage they cause [12]. If the conditions are not ideal, it takes 30 to 40 days for RKNs to produce a new generation [12]. The developmental stages involved in the life cycle of RKNs are the egg stage, four juvenile stages (J1, J2, J3, and J4), and the adult stage. The first juvenile stage, J1, molts within the egg and undergoes a second molt, which results in the emergence of the J2 stage from the egg. The J2 stage is the infective stage of RKNs and the only motile stage for these sedentary endoparasitic nematodes. After hatching, the juvenile moves towards the plant roots and penetrates a root tip using its stylet and enzymes produced by its secretory glands [12, 13]. Following penetration, the juvenile moves along the cells (intercellular movement) to the zone of cell differentiation, where they become sedentary and form giant feeding cells, which are their permanent feeding sites [12]. Here, the J2s molt into J3s and J4s and finally into an adult male or female. After mating, the adult male moves out of the plant into the soil where they die while the female develops an ovoid shape and lays eggs that give rise to another generation.

2.1 Plants susceptible to attack by root-knot nematodes

Various plants are susceptible to attack by root-knot nematodes because they are polyphagous microorganisms. The host plants for *Meloidogyne* spp. include tomatoes, finger millet, cucumbers, eggplants, peas, cotton, cowpeas, coffee, okra, pawpaw, sugar beet, sunflowers, tobacco, potatoes, beans, and peppers [12–16]. RKNs also have a complex relationship with Striga weeds, which are plant-parasitic weeds that compete with crops in farms for nutrients and therefore affect crop yields [17]. Root-knot nematodes can affect these plants during different developmental stages including the seedling phase, vegetative growth phase, and maturation phase [12]. Further, the RKNs affect the whole plant, its roots, or leaves, producing observable symptoms. Conditions that favor the development of root-knot nematodes include warm temperatures, moist and well-aerated soils, planting monocultures, growing susceptible crops in infested soils continuously, and weeds in fields [12].

2.2 Symptoms associated with root-knot nematode attack

RKNs produce aboveground and belowground symptoms in affected plants [7, 12]. The aboveground symptoms of nematode infestation include yellowing of leaves, patchy fields and stunted growth while belowground symptoms include galled, swollen, or distorted roots, reduced root volume, and stunted root growth. The diameter of the galls (root knots) ranges from smaller than a pinhead to 25 mm [12]. According to the plant parts affected, root-knot nematode infestation produces the following symptoms: wilting in leaves, galling, swelling, distortion, or reduced volume in roots, and dwarfing or wilting for the whole plant. Since these symptoms such as wilting and yellowing leaves might be similar to symptoms produced by other biotic or abiotic factors, it is important for farmers to contact specialists for accurate diagnosis of their crops, to avoid further damage by plant-parasitic nematodes, and to implement effective and long-term corrective measures [7, 8, 12].

2.3 Methods for managing root-knot nematodes

While the most reliable measures for controlling RKNs are preventive, management measures exist. Farmers can use different strategies for the management of root-knot nematodes. These include crop rotation, planting resistant varieties, use of plant extracts that suppress the activity of RKNs, use of trap crops, sanitation, and use of chemical nematicides [7, 8, 18]. These methods enable farmers to reduce existing RKN populations and are suitable for seasonal crops and the establishment of woody plants [12]. Even though these methods are effective, they are short-term solutions to RKN infestations and mostly reduce their populations in the top layer of soil. Additionally, since *Meloidogyne* spp. have a wide host range; the effectiveness of cultural practices becomes reduced.

Mostly, farmers use pesticides to control the nematodes. While some of these pesticides are effective, they are non-specific and therefore harm non-pathogenic microorganisms in the environment [3]. Additionally, due to the chemical nature of the pesticides, they act as environmental pollutants with a high residue effect. Since biocontrol confers several benefits to the farmers, the crops, and the environment, it is the most suitable method to use for the inhibition of the parasitic activities of *M. incognita*. Plant extracts from *Eucalyptus citriodora* (eucalyptus), *Azadirachta indica* (neem), and *Tagetes erecta* (marigold) are effective against *M. incognita*, *Hoplolaimus*, and *Helicotylenchus multicinctus* [6]. Farmers can also use nematophagous fungi and bacteria in the biocontrol of *M. incognita*. Parasitic bacteria belonging to the *Pasteuria* and *Bacillus* genera are effective against *M. incognita* [6]. Nematophagus fungi like *Lecanicillium psalliotae*, *Pochonia chlamydosporia*, and *Trichoderma* spp. are also effective biocontrol agents [6, 19].

3. *Trichoderma* spp.

Trichoderma spp. are ubiquitous and saprotrophic microorganisms classified as Ascomycetes [15, 20]. One can isolate these fungi from decomposing wood, soil, and other organic materials from plants. The scientific classification of the fungi identifies it as imperfect fungi because it has no known sexual (haploid) stage [21]. Asexual reproduction of the fungi occurs through the production of conidiospores. In culture, the fungi exhibit a rapid growth rate and produces numerous spores that have different shades of green at maturity (**Figures 1–3**) [21]. On the reverse side of a culture plate, *Trichoderma* spp. appears uncoloured, yellow, yellow-green, or amber [21]. The biological control and plant promotion properties of different strains in this genus have generated significant scientific interest, resulting in numerous studies since the discovery of the microbe. Additionally, the potential of *Trichoderma* spp. resulted in industries producing the fungi commercially.

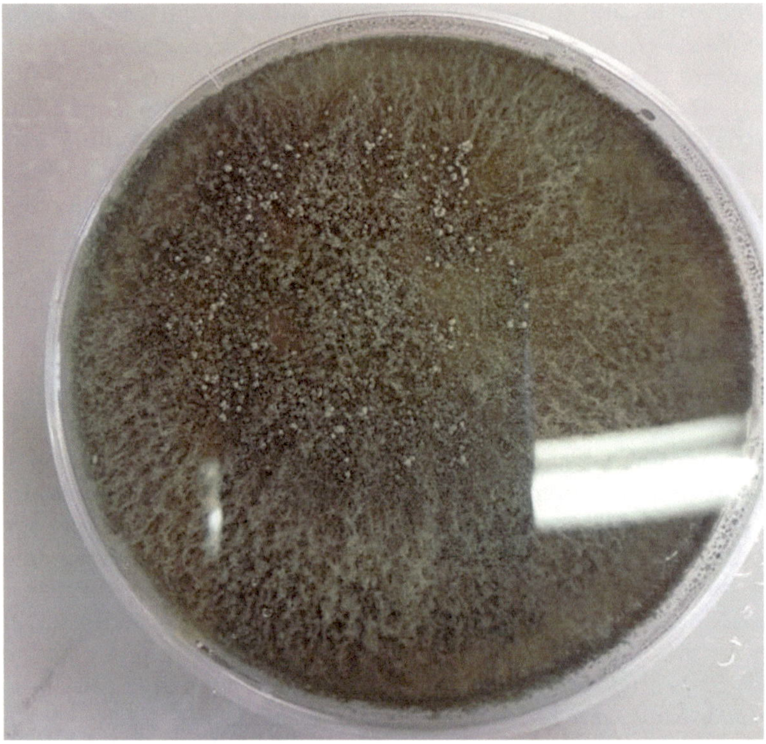

Figure 1.
Trichoderma asperellum *growth on a PDA plate.*

Figure 2.
Trichoderma atroviride *growth on a PDA plate. The yellow or cream growth represents immature spores while the green growth represents mature spores.*

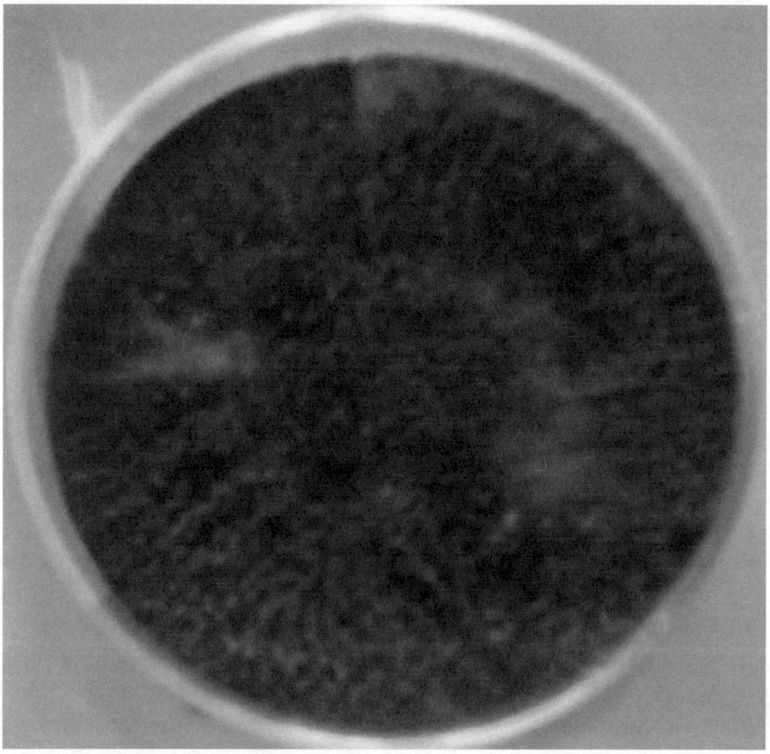

Figure 3.
Trichoderma hamatum growth on a PDA plate.

3.1 *Trichoderma* spp. in plant growth promotion

Scientific studies on the interactions between plants and *Trichoderma* spp. have made it possible to understand how *Trichoderma* spp. promote plant growth [18–24]. Additionally, these studies show that selecting efficient strains of the fungi is crucial because the efficient strains develop a chemical communication with the host plant after infecting and colonizing the outer layers of its plant roots [22]. Initially, the infection and colonization produce host plant resistance mechanisms against the fungi that inhibit further colonization in most plants. However, in some plants like the cocoa plant, the response allows the fungi to ramify the plant structure [23]. Several chemical factors including peptides, hydrophobin-like proteins, and smaller molecules initiate plant responses [23]. While some scientists continue to discover these factors, others have documented the identified active metabolites [23]. Following colonization and development of chemical communication by efficient strains, *Trichoderma* spp. persist in association with the host plant and confer several benefits to the plant throughout its lifecycle.

Trichoderma spp. are endophytic plant symbionts that can multiply and grow in soil [23, 24]. The root colonization and subsequent establishment of chemical communication with the host plant result in an alteration of the plant's physiology due to a change in the plant's genetic expression [8, 9, 11, 14, 20, 23]. The change affects several known plant genes or proteins and generally enhance a plant's performance. Remarkably, although the fungal containment is in the roots, there is a greater change in gene expression in the shoots compared to the roots. One advantage of root colonization by *Trichoderma* spp. is improved systemic resistance against plant diseases through the plant's innate defense mechanisms [9, 11]. Activating the innate defenses reduces the diseases caused by many pathogenic microorganisms including bacteria, fungi, and viruses. The induced systemic resistance occurs in both monocotyledonous and dicotyledonous plants [11]. Even though innate defense mechanisms are not fully effective and rarely approach immunity, *Trichoderma* spp. offsets this setback by inducing long-term resistance, since the induced protection persists for months after application, and provides integrated control for pathogenic microorganisms because *Trichoderma* sp. show resistance to most chemical pesticides [11]. Commercial seed treatment systems demonstrate that a good approach for integrated control involves the application of a chemical pesticide with effective *Trichoderma* spp. [11]. By applying these two control agents, plants have increased pathogen protection because pesticides offer comparatively short-term but more effective pathogen control while *Trichoderma* spp. offers long-term protection.

Apart from protection, some *Trichoderma* strains confer other benefits to plants by enhancing the plant's resistance to abiotic stresses such as unfavorable temperature and water and nutrient unavailability [11]. A principal mechanism that *Trichoderma* spp. uses to offer resistance to abiotic stress is enhancing the expression of the enzymes involved in foraging reactive oxygen species [11]. Plant exposure to biotic or abiotic stressors leads to the production of destructive amounts of reactive oxygen species and some symptoms of this exposure result from this toxicity. Enhancing enzymes in pathways like the glutathione-ascorbate cycle improves the speed with which antioxidant recycling occurs, reducing the effects of stress exposure [11]. Further, *Trichoderma* spp. induce better nitrogen use efficiency in plants. In research to identify the amount of applied nitrogen fertilizer taken up by plants, the researchers found that plants take up only 33% of the fertilizer, which is a low quantity [11]. Further research in this area shows that it is possible to reduce the application of nitrogen fertilizer by 30–50% without affecting the yield [11]. Even though research has not fully discovered the mechanisms involved in the induction of nitrogen use efficiency in plants, one factor that contributes to this phenomenon is *Trichoderma*'s ability to increase the depth of a plant's root system. Even though the effects triggered by *Trichoderma* infection require energy and would therefore lead to reduced plant growth, scientists continuously observe increased growth [11]. An explanation offered for this observation is that *Trichoderma* spp. can improve a plant's photosynthetic efficiency, which has significant implications in plant growth given that plants depend on sunlight and photosynthesis for energy [23].

Newer strains of *Trichoderma* with the discovered beneficial characteristics have significant advantages for farmers in developed and developing countries because first, since they grow together with plant roots, the farmers need only a small amount of inoculum to realize the fungi's long-term benefits [11]. Developing countries can source their inoculum from developed countries, which produce the microbe under high quality and controlled environments. The developed transaction relationship is economical because the shipping of small inoculum amounts is sufficient and since developing countries might not have the resources required to produce pure isolates for inoculation, sourcing them from developed countries would be more logical.

Effective induction of plants to cope with biotic and abiotic stressors has desirable environmental implications that include reduced water and air pollution due to the production and use of nitrogen fertilizers. Additionally, farmers in developed and developing countries experience improved yields, which gives them a good return on their investments. The improved yields being the cumulative effect of a plant's improved photosynthetic efficiency and responses to biotic and abiotic stressors [11, 23]. Apart from improved yields, the reduced application of nitrogen fertilizers improves farmers' savings. In developing countries, since farmers with small farms might not afford commercial fertilizers and pesticides, applying *Trichoderma* spp. that cost comparatively less to improve plant growth can contribute significantly to global food security. Due to the progressive increase in the global population, these improved yields are crucial to preventing a possible food shortage crisis. *Trichoderma* spp. fungi are significant in plant growth promotion because they are easy to manipulate or select since it is possible to grow and select several strains using simple techniques. Additionally, the fungi have a reasonably long shelf life that is important in commercialization [11]. However, since the benefits of *Trichoderma* spp. are strain-specific, it is impossible to generalize the effects of one strain to others. Consequently, with the discovery of a new strain, researchers should study its plant growth characteristics.

3.2 *Trichoderma* spp. as a biocontrol agent

Different species within the *Trichoderma* genus exhibit different biological control mechanisms; consequently, to apply a *Trichoderma* spp. fungus efficiently and effectively, scientists and farmers need to understand its mode of action and limitations.

3.2.1 Mycoparasitism and antibiotic (toxin) production

A salient characteristic of members of *Trichoderma* spp. as research indicates is the ability to parasitize fungi in other genera [8, 21]. Mycoparasitism is an obligate mode of nutrition for mycoparasitic fungi, for example, *T. lignorum* because despite an external supply of essential nutrients the fungi still parasitizes other fungi [21]. The biocontrol mechanisms exhibited by mycoparasitic fungi are coiling around the pathogen's hyphae, penetrating the hyphae, and dissolving the pathogen's cytoplasm. Later, research showed that a *T. lignorum* strain produces toxic substances into its environment to facilitate mycoparasitism [21]. The substance produced is toxic to *Rhizoctonia solani* and *Sclerotinia americana.* The researcher who discovered this toxic substance named it gliotoxin. Following this discovery, researchers show that it was *Gliocladium virens*, later renamed *Trichoderma virens* that produces gliotoxin [21]. After this discovery, subsequent studies ascribed effective biocontrol by *Trichoderma* spp. to antibiosis and mycoparasitism [21]. The antibiotic activity of different fungi in the *Trichoderma* genus is specific to some microorganisms. For example, gliovirin produced by *T. virens* is effective against *Pythium ultimum* and *Phytophthora* sp. but not against several other microorganisms including the bacteria *Bacillus thuringesis, Pseudomonas fluorescens,* and the fungi *R. solani* and *Rhizopus arrhizus* [21]. Other *Trichoderma* spp. that produce toxic substances against pathogenic microorganisms are *T. koningii* strain T-8 and *T. harzanium* strain T-12. Later, research showed that *T. virens* strains deficient in toxin and mycoparasitism genes were still effective biocontrol agents, which gave rise to the concept of competition and rhizosphere competence as a biocontrol mechanism for *Trichoderma* spp. [21].

3.2.2 Enzymes

Recent research shows iterative results for another mechanism used by *Trichoderma* spp. for biocontrol [21]. Based on these results, fungi in the genus produce chitinases and/or glucanases, which are enzymes that hinder the development of plant pathogens [8, 15, 21]. The mode of action of these enzymes involves breaking down the glucans, polysaccharides, and chitin that confer rigidity to the cell walls of pathogenic microorganisms such as fungi and plant-parasitic nematodes. Consequently, *Trichoderma* spp. interfere with the cell wall integrity of the pathogens. Further research that involved disrupting or over-expressing the genes that code for chitinase showed mixed biocontrol activity by *Trichoderma* spp., with the transformant fungi showing increased or decreased biocontrol activity towards select pathogens [21]. Due to these results, Scientists concluded that other factors and mechanisms apart from chitinases are responsible for the biocontrol process [21]. Still, in research to determine the role of chitinases in biocontrol, scientist created transgenic plants by transferring the genes that encode endo and exochitinases from *Trichoderma* spp. to tobacco, potato, apple, and cotton plants [21]. The genetically modified crops showed increased resistance against plant pathogens compared to the non-transgenic lines, proving the role of chitinases in biocontrol by *Trichoderma* spp. fungi. Further, research shows that *Trichoderma* spp. produce protease enzymes that inactivate the hydrolytic enzymes produced by pathogenic microorganisms by breaking down the pathogen's enzymes into their precursor molecules, destroying their ability to infect plant cells [21]. Consequently, *Trichoderma* spp., specifically *Trichoderma harzanium*, reduce the severity of diseases caused by root-knot nematodes and fungal pathogens through protease activity. Further research on the protease activity of *T. harzanium* in biocontrol used a transformed strain of the fungus and the results showed that the transformed strain was more effective than the wild type strain in reducing root galls due to root-knot nematode infestation [21]. Additionally, the transformed strain exhibited a unique trait by penetrating the egg masses and the eggs inside the galls, hence reducing their pathogenicity. In advanced studies, researchers used a combination of antibiotics and enzymes to illustrate the effects of synergizing two biocontrol mechanisms [21]. Based on the results of these studies, combining antibiotics and enzymes produced superior results in terms of reducing pathogenic activity compared to each treatment alone. However, the effective application of the effects of synergism depends on a better understanding of the components in the association.

3.2.3 Induction of defense responses in plants

Studies also propose that the biocontrol activity of *Trichoderma* spp. is due to the fungi's ability to induce resistance in their host plant [9, 24]. In studies using *Trichoderma harzanium* as the biocontrol agent, researchers demonstrated that specific concentrations of the fungus initiated defense responses in the roots and leaves of the host plants [21]. The plant responses involved enhanced chitinase activity, increased peroxidase activity, which causes the production of compounds that are toxic to fungi and deposition of callose-enriched wall appositions on inner cell wall surfaces. Since *Trichoderma* spp. are more resistant to antifungal plant responses compared to pathogenic fungi, the association between the plant roots and the fungi results in the formation of a symbiotic mycorrhizal relationship [21]. On the other hand, the enhanced host plant defense responses are lethal to pathogenic microbes, hence their death.

3.2.4 Adjunct mechanisms

Even though these additional mechanisms are not primary biocontrol mechanisms, they promote disease tolerance or resistance in the host plants [8, 21]. The manifestation of these characteristics includes increased root and shoot growth in plants, changes in a plant's nutritional status, and a plant's increased resistance or tolerance to biotic and abiotic stresses. For example, treating plants in nitrogen-deficient soils with *Trichoderma* spp. results in healthier plants with improved yields compared to untreated crops [10, 21]. Research shows that this effect could result from the symbiotic interactions between *T. harzanium* and *Bradyrhizobium japonicum*, a nitrogen-fixing bacterium. Theoretically, the association between these two microorganisms improves the plant's nitrogen utilization capacity, which decreases the need to use artificial nitrogen fertilizers on a farm [21]. Further, research shows that plants treated with *T. harzanium* have improved nutrient concentrations due to the beneficial interaction with *Trichoderma* spp.

3.3 *Trichoderma* spp. as a biocontrol agent for root-knot nematodes

Results from previous studies indicate that different *Trichoderma* spp. including *T. atroviride, T. viride, T. asperellum, and T. harzanium* are excellent biocontrol agents against root-knot nematodes [10, 19–24]. Using *Trichoderma* spp. when growing plants susceptible to root-knot nematodes reduces the formation of root galls due to nematode infestation and promotes plant tolerance and growth [8, 14, 18, 20]. *Trichoderma* spp. have highly branched conidiophores that produce conidia, which attach to various nematode stages. The attachment and parasitic activity of these conidia depend on the *Trichoderma* species and strain; however, successful parasitism of root-knot nematodes during any stage requires mechanisms that promote penetration of pathogen eggs and cuticles by the antagonistic *Trichoderma* spp. [14]. Research on the attachment of these fungi shows that it results from the formation of appressoria due to fungal coiling. Further research shows that lytic enzymes for example β-1, 3-glucanase, chitinase, protease and lipase produced by *Trichoderma* spp. have a role in *Meloidogyne* spp. parasitism [14, 15]. Apart from direct antagonism, *Trichoderma* spp. use other techniques including induced plant resistance and fungal metabolites discussed earlier that are useful in the biological control of *Meloidogyne* spp. In research to determine the modulation of the hormone-signaling network of the host plant by *Trichoderma* spp. for induction of nematode resistance, the researchers observed that plant roots colonized by the fungi hindered RKN development locally and systematically [25]. The hindrance occurred at different stages including the reproductive, gall formation, and penetration stages. The fungi achieved this effect by priming defenses regulated by salicyclic acid, which prevents root invasion by J2s [24]. They also boosted the plant defenses controlled by jasmonic acid, which prevented the nematodes from provoking the deregulation of jasmonic acid-dependent immune responses that hindered the formation of root galls and female fertility [24].

Overall, to receive optimal outcomes, farmers and researchers should apply *Trichoderma* spp. to the soil before planting to promote the proper establishment of the fungi on the plant rhizosphere, which is crucial to the management and control of root-knot nematodes. Augmentative biological control describes the process of applying selected and mass-produced biological control agents such as *Trichoderma* spp. fungi in high densities one or many times during a planting season.

4. Conclusion

Agriculture is an important economic activity globally because it is a source of income for many families and contributes substantially to the global gross domestic product. Crop farmers engage in both subsistence and cash crop farming to sustain the ever-growing global population. Consistent with the population growth, farmers adopt other farming practices including using more land to increase their productivity. However, the pests and diseases that attack crops in the field challenge this anticipation. Pests including rodents, birds, and insects cause direct and indirect damage to crops through their activity. Insects like aphids, for example, cause direct damage by sucking nutrients from the phloem of plant tissues and indirectly by acting as vectors for pathogenic microbes, which can enter plant tissues as the insect feeds and multiply leading to plant diseases. Plant diseases arise from abiotic, mesobiotic, and biotic factors. While abiotic factors are non-living components of a plant's ecosystem, mesobiotic factors exhibit an intermediary state between a living and non-living organism, for example, virus particles. Biotic factors that cause plant diseases are prokaryotic and eukaryotic organisms in a plant's environment. The prokaryotic organisms are bacterial cells and eukaryotic organisms include fungi and nematodes. Plant-parasitic nematodes, especially root-knot nematodes, are among the most significant biotic pathogens worldwide because they cause significant losses to farmers of different food types. Controlling root-knot nematodes is usually difficult because they are polyphagous and can form dauer stages, which enable them to survive in the soil for long periods until they detect a susceptible host plant. However, biocontrol agents such as *Trichoderma* spp. fungi provide long-term and effective solutions against RKNs. Additionally, the fungi promote plant growth in the plants it colonizes, hence a desirable double effect. However, it is crucial to understand the mechanism of action of the various fungi in the *Trichoderma* genus for their efficient application.

Acknowledgements

I would like to acknowledge my undergraduate supervisors, Ms. Beth Wangui Waweru and Dr. Njira Njira Pili, who spurred my interest in biological control agents for root-knot nematodes during my project, pointed me to quality resources for the subject matter and helped improve the quality of my work.

Conflict of interest

The author declares no conflict of interest.

Thanks

My gratitude goes to Njagi for encouraging me to write this chapter; Judah Leo, who inspires me to try out new opportunities, achieve greater heights and give my best; my parents, Mr. and Mrs. Mulusa, for supporting my academic goals; and IntechOpen for this opportunity.

Nomenclature

1. Abiotic factors – These are non-living components of an ecosystem that influence the ecosystem. Examples of abiotic factors are light, water, humidity, acidity, and temperature.

2. Biocontrol - Biological control of plant diseases involves using living organisms to suppress plant pathogen populations.

3. Biocontrol agents - Scientists and farmers apply microbial biological control agents (MBCAs) to crops for the biological control of plant pathogens. The MBCAs use various modes of action to control the pathogens effectively. Some of them induce plant resistance and control the pathogen without any direct MBCA-pathogen interactions; others compete for nutrients with pathogens or exhibit other mechanisms that modulate the pathogen growth conditions; and others use antagonism through hyperparasitism and antibiosis, which inhibit the pathogens directly. Metabolic events that combine various modes of action regulate these interactions.

   It is crucial to understand an MBCA's mode of action to optimize its pathogen-control abilities. Additionally, understanding the mode of action promotes the succinct knowledge and characterization of risk exposure for organisms higher up in the food chain and the environment as well as plant resistance development towards a given MBCA.

4. Biotic factors – Biotic factors are living components of an ecosystem that affect other living organisms within the ecosystem or the ecosystem as a whole. In a farm, the biotic factors can include insects, microorganisms, plants, and rodents.

5. Dwarfing – This is a process that occurs when there is a change in a plant so that its size is significantly smaller than that of other plants within the same species. The causes of dwarfing can be hormonal, genetic, or nutritional.

6. Fungus – A type of eukaryotic living organism with a filamentous, unicellular, or multicellular existence. The scientific name used to refer to the filament is hyphae (plural: hypha). The cells of fungi have chitinous or cellulose cell walls and while some fungi are parasitic, others are saprophytic. Further, fungi can produce either sexually or asexually.

7. Nematophagus fungi – These are fungi that trap, kill, and digest nematodes using specialized structures on their mycelia or spores for trapping vermiform nematodes or hyphal tips for attacking nematode eggs and cysts before penetrating the cuticles of nematodes.

8. Nematicides – These are chemical pesticides used for killing plant-parasitic nematodes. Often, they are broad-spectrum toxicants with high volatility, which facilitates their movement in soil following application. An example is fosthiazate.

9. Pesticides – These are substances or a mixture of substances used to prevent, terminate, repel, or diminish pests. Application methods for pesticides include spraying, dusting, padding, granular application, and seed pelleting.

10. Plant extracts – These are substances with desirable properties drawn from a plant tissue for a specific use. Often extraction involves the use of solvents such as ethanol.

11. Pollutants – These are contaminants that when introduced to the environment, affect it adversely. They include particulate matter, greenhouse gases, and chemicals.

12. Reactive oxygen species – These highly reactive chemical molecules result from oxygen's electron receptivity. Examples are alpha-oxygen and peroxides.

13. Root galls – Unusual swellings or localized tumors in plant tissues. Often, their size varies, ranging between 1 and 10 mm in diameter. The size depends on the nematode species and the location of the gall on the root system. Severe galling causes root malformation, shortening, and thickening, which hinders development and branching in roots.

14. Striga weeds – Commonly known as witchweed, this parasitic crop occurs naturally in parts of Australia, Africa, and Asia. The weed parasitizes cereal crops mostly, which reduces their yield significantly.

15. Wilting – A phenomenon observed when non-woody plants lose rigidity due to a decrease in the turgor pressure of non-lignified plant cells. Wilting occurs due to reduced water supply to the cells.

Author details

Linnley Mulusa
Moi University, Eldoret, Kenya

*Address all correspondence to: djouneywilley@gmail.com

References

[1] The World Bank. Agriculture and Food [Internet]. 2021. Available from: https://www.worldbank.org/en/topic/agriculture/overview#1

[2] M-Farm. Potato production in Kenya [Internet]. 2014. Available from: https://mfarm.co.ke/blog/post/potato-production-in-kenya

[3] Fauster Admin. Crop Pests and Diseases - agriculture form 3 notes [Internet]. 2020. Available from: https://www.easyelimu.com/high-school-notes/agriculture/form-3/item/2086-crop-pests-and-diseases

[4] Tjosvold S and Koiko, S. Distinguishing Abiotic And Biotic Symptoms In Plants [Internet]. 2015. Available from: https://ucanr.edu/blogs/blogcore/postdetail.cfm?postnum=17827

[5] University of California. Nematodes/Potato/Agriculture: Pest Management Guidelines/UC Statewide IPM Program (UC IPM) [Internet]. N.d. Available from: https://www 2.ipm.ucanr.edu/agriculture/potato/nematodes/

[6] Bernard GC, Egnin, M. and Bonsi, C. The impact of plant-parasitic nematodes on agriculture and methods of control. Nematology – concepts, diagnosis and control. 2017 Aug 16; 121

[7] Stirling GR, Nicol J, Reay F. Advisory services for nematode pests. Rural Industries Research and Development Corp.; 1999.

[8] Poveda J, Abril-Urias P, Escobar C. Biological control of plant-parasitic nematodes by filamentous fungi inducers of resistance: Trichoderma, mycorrhizal and endophytic fungi. Frontiers in Microbiology. 2020;11. DOI: 10.3389/fmicb.2020.00992

[9] Forghani F, Hajihassani A. Recent advances in the development of environmentally benign treatments to control root-knot nematodes. Frontiers in Plant Science. 2020;11. DOI: 10.3389/fpls.2020.01125

[10] Harman GE. Trichoderma—not just for biocontrol anymore. Phytoparasitica. 2011 Apr 1;39(2):103-108. DOI: 10.1007/s12600-011-0151-y

[11] Abad P, Favery B, Rosso MN, Castagnone-Sereno P. Root-knot nematode parasitism and host response: molecular basis of a sophisticated interaction. Molecular plant pathology. 2003 Jul;4(4):217-224. DOI: 10.1046/j.1364-3703.2003.00170.x

[12] Infonet Biodivision. Root-knot nematodes [Internet]. N.d. Available from:https://infonet-biodivision.org/PlantHealth/Pests/Root-knot-nematodes

[13] Przybylska A, Obrępalska-Stęplowska A. Plant defense responses in monocotyledonous and dicotyledonous host plants during root-knot nematode infection. Plant and Soil. 2020 Apr 29:1-22. DOI: 10.1007/s11104-020-04533-0

[14] Mukhtar T. Management of root-knot nematode, Meloidogyne incognita, in tomato with two Trichoderma species. Pakistan Journal of Zoology. 2018 Aug 1;50(4). DOI: 10.17582/journal.pjz/2018.50.4.sc15

[15] Sayed M, Abdel-rahman T, Ragab A, Abdellatif A. Biocontrol of Root-Knot Nematode Meloidogyne incognita by Chitinolytic Trichoderma spp. Egyptian Journal of Agronematology. 2019 Jan 1;18(1):30-47. DOI: 10.21608/ejaj.2019.52842

[16] NC State. Control of root-knot nematodes in the home vegetable garden [Internet]. 2018. Available from: https://content.ces.ncsu.edu/

contol-of-root-knot-nematodes-in-the-home-vegetable-garden#:~text=Other%20common%20garden%20vegetables%20grown,have%20galled%20and%20decayed%20roots.

[17] Fernandez-Aparicio M, Delavault P, Timko MP. Management of infection by parasitic weeds: A review. Plants. 2020 Sep;9(9):1184. DOI: 10.3390/plants9091184

[18] Fan H, Yao M, Wang H, Zhao D, Zhu X, Wang Y, Liu X, Duan Y, Chen L. Isolation and effect of Trichoderma citrinoviride Snef1910 for the biological control of root-knot nematode, Meloidogyne incognita. BMC microbiology. 2020 Dec;20(1):1-1. DOI: 10.1186/s12866-020-01984-4

[19] Al-Hazmi AS, TariqJaveed M. Effects of different inoculum densities of Trichoderma harzianum and Trichoderma viride against Meloidogyne javanica on tomato. Saudi Journal of Biological Sciences. 2016 Mar 1;23(2):288-292. DOI: 10.1016/j.sjbs.2015.04.007

[20] Studholme DJ, Harris B, Le Cocq K, Winsbury R, Perera V, Ryder L, Beale M, Ward J, Thornton CR, Grant M. Investigating the beneficial traits of Trichoderma hamatum GD12 for sustainable agriculture—insights from genomics. Frontiers in plant science. 2013 Jul 30;4:258. DOI: 10.3389/fpls.2013.00258

[21] Howell CR. Mechanisms employed by Trichoderma species in the biological control of plant diseases: the history and evolution of current concepts. Plant disease. 2003 Jan;87(1):4-10. DOI: 10.1094/PDIS.2003.87.1.4

[22] Köhl J, Kolnaar R, Ravensberg WJ. Mode of action of microbial biological control agents against plant diseases: relevance beyond efficacy. Frontiers in Plant Science. 2019 Jul 19;10:845. DOI: 10.3389/fpls.2019.00845

[23] Harman GE, Doni F, Khadka RB, Uphoff N. Endophytic strains of Trichoderma increase plants' photosynthetic capability. Journal of applied microbiology. 2021 Feb;130(2):529-546. DOI: 10.1111/jam.14368

[24] Molinari S, Leonetti P. Bio-control agents activate plant immune response and prime susceptible tomato against root-knot nematodes. PloS one. 2019 Dec 3;14(12):e0213230. DOI: 10.1371/journal.pone.0213230

[25] Martinez-Medina A, Fernandez I, Sanchez-Guzman MJ, Jung SC, Pascual JA, Pozo MJ. Deciphering the hormonal signaling network behind the systemic resistance induced by Trichoderma harzanium in tomato. Frontiers in Plant Science. 2013 Jun 24;4:206. DOI: 10.3389/fpls.2013.00206

Section 4

Nematodes Biological Indicators

Chapter 10

Nematodes as Biological Indicators of Soil Quality in the Agroecosystems

Tabassum Ara Khanum, Nasir Mehmood and Nasira Khatoon

Abstract

Soil nematodes have advantages as bio-indicators, because they have beneficial role in the food web. Nematodes associated with bacteria are probably the most studied biological indicators of soil fertility. Saprophytic nematodes act as bio-indicators of soil health because they have different beneficial ways to increase in soil functions such as in management of ecosystem; enhancement of nitrogen in soil by ingestion of nitrogen and secrete extra nitrogen as NH4, that is easily absorbable; putrefaction and by dispersion of bacteria and fungi to recently available organic residues. Therefore, nematode are beneficial in increasing soil health or plant growth by providing the nutrient through associated bacteria. So it can be evaluated that the nematodes use as biological indicators of soil fertility because of remarkable diversity and nematode contribution in many functions of the soil fertility.

Keywords: Soil health, indicator, microorganism, nematodes, diversity, food web, nitrogen mineralization

1. Introduction

Biological indicators have association with different functions of soil and have quality to monitor the soil functions and enhance the health of soil [1, 2]. These indicators play a dynamic role to increase the soil properties, in decomposition and chemical contaminants. Many scientists [3] and Pankhurst *et al.*, [4] Haitova and Bileva [5] preferred that biological indicator, indicates ecosystem processes; physical, chemical changes and biological properties and processes [6, 7]. Soil health is the quality of a soil to function within ecosystem's limits to sustain in biotic activity [8], to keep environmental quality for the promotion of plant and animal health. A healthy soil will be needed to help in life processing such as plant production and support soil food web and maintain microbial diversity Vads *et al.*, [9]. In agrarian countries the use of chemical fertilizers make independent in food production but it makes polluted atmosphere and cause hazardous impacts on anima and humans. Due to inadequate uptake of these chemical based fertilizer by plants, they absorb into water bodies through rainy water, by which water bodies enriched with nutrients and minerals and effect on biotic fauna flora and also the growth living microorganism. The extra uses of chemical fertilizers in agrarian fauna are more expensive and also have several antagonistic effects on soils as reduction of water holding capacity, soil fertility and disparity in soil nutrients.

Soil nematodes have been recognized as the part of agrarian fauna as they have a significant role in the ecosystem [10–12]; Bileva and Arnaudova, [13]. Nematodes as bio-indicators have been played a key roles in the decomposition of soil organic matter, food chain cycling, degradation of soil pollution, and the formation of healthy soil structure [14]. They have the capability to make differences in their areas, such as stress due to deficiency and contaminants. Biological indicators may reproduce the overall population, category, and activity of microorganisms and the diversity of the living organisms in soil [15].

The presence of plant growth promoting rhizobacteria (PGPR) in soil, is also biological indicator because the microorganisms rebuild the nutrient cycle and maintain the organic matter in the soil. Through the use of PGPR, vital plants can be grown in the soil. Since they play several roles, a preferred scientific term for such beneficial bacteria such as *B. cereus, B. subtilis* [16] and some species of *Serratia* provide "eco-friendly" organic agro-input. Soil nematode of the family Rhabditidae are associated with different bacteria and the secondary metabolites produced by these bacteria have the ability to fix nitrogen in the soil. *Oscheius* is an excellent laboratory model to study internal gene transfer because these microbial worms are bacterial feeders, have vector viable bacteria *B. cereus, B. megaterium, B. subtilis* and *Pseudomonas aeruginosa*. They have the ability to absorb PO4. Insoluble PO4 is generally un-absorbable to the plant. The root system simply cannot absorb it. Soil microorganisms, such as *B. subtilis*, are beneficial to plant health and plant growth.

Besides saprophytic nematodes, free-living marine nematodes use as pollution indicators coastal areas. The use of benthic flora as bio-indicators of different water source like, ocean, river and lake quality can be examined in terms of population density and diversity, test morphology - including size, prolocular morphology, ultrastructure, abnormality, and the chemistry of the test. The study of pollution effects on benthic flora and their use as substitutions began in the 1960s [17–19], and has been increasingly developed in recent decades as a result of environmental research (for reviews, see [20–25]).

Soko, and Gyedu-Ababio [26] reported the relationship between different environmental factors and with free-living marine nematodes. They found some metals such Cadmium, Colbat, Chromium, Copper, Iron, Manganese, Nickel, Vadium, Zinc and Aluminum affected the diversity and density of marine nematodes. Shannon-Wiener Diversity, Maturity Index and colonize-persisters percentage (c% - p%) were also found to be good tools for use as pollution indicators Chander & Brookes, [27]. Nematode genera such as *Terschellingia*, *Theristus* and *Halalaimus* were found during that study to be dominant at a site strongly impacted by both metals concentration and organic matters. The three genera are believed to be good indicators of pollution in the Incomati River Estuary.

2. Materials and methods

2.1 Soil sampling and nematode isolation

Surveys were conducted to check the soil fertility and nematode presence. The experiment was laid out in randomized complete block design (RCBD). The trial was conducted and was repeated to evaluate the nitrogen mineralization and presence of saprophytic nematodes associated with bacteria. In this experiment 5 soils samples (each field) were collected from two different types of vegetation, one from healthy plantations and another from infected plants.

Soil samples consisting of 1.6 cm diam., x 10 cm deep cores were taken from each plot. The samples of the same plot were mixed thoroughly to form a composite sample. 100 g soil samples taken from each composite sample were processed by

Cobb sieving [28] followed by modified Baermann funnel method [29]. Nematodes were collected after 48 h. The population of saprophytic nematodes was increased in the presence of symbiotic bacteria. Plant growth was also increased so it indicates that nematodes use as bio-indicator of plant and soil health.

2.2 Soil analysis

In the present experiment sandy loam soil was used, which was collected from Botanical garden of University of University of Karachi. Soil analysis was performed in the Department of Environmental studies.

3. Results and discussion

In this experiment 5 soils samples (each field) were collected from two different types of vegetation, one from healthy plantations and another from infected plants. Saprophytic and parasitic nematodes were the most abundant groups in all samples. There were significant differences in the numbers of saprophytic ($P \leq 0.01$) and plant parasitic nematodes ($P \leq 0.05$) between the healthy and infected plantations but no difference was observed in the numbers of fungal feeding nematodes. The common genera were found in all samples (*Aphelenchoides*, *Aphelenchus*, *Meloidogyne*, *Pratylenchus*, *Trichodorous*, *Helicotylenchus*, *Hoplolaimus*, *Xiphinema*, *Tylenchus* and *Mononchus*. From healthy plantation the bacteria feeding genera, *Acrobeles*, *Rhabditis*, *Cervidellus*, *Eucephalobus*, *Cephalobus*, *Heterocephalobus*, *Plectus* and *Tylocephalus* were found, showed that these nematodes fixed the nitrogen fixing bacteria in the soil for which soil is healthy source for healthy plantation or it can be evaluated that the presence of nematodes indicates the soil health [30, 31].

3.1 Soil analysis

The soil composed of 55% of sand, 27.4% silt and 16.5% clay and contained PO4 2.5 mg/kg, total N: 11.42 mg /kg by TKN method [32] and the pH was 7.1. Fresh soil passed through a 2 mm mesh to remove stones, macro-fauna and discernible. The half of soil was sterilized, which contained PO4 2.4 mg/kg, total N: 103 mg /kg by TKN method (Total Kjeldhal Nitrogen) and the pH was 6.8.

Figure 1.
Healthy plants roots, stem and leaves showed the fertility of soil.

Figure 2.
Infected plants A, B and C and healthy plants D and E.

The saprophytic nematode population significantly increased in healthy soil (RCBD- ANOVA: F = 17; df = 8, 18; P < 0.001) **Figure 1**. Plant parasitic and free-living soil nematodes also differed significantly (RCBD - ANOVA F = 25, df = 2, 18; P < 0.001) and the presence or absence of plant and soil also had marked effect on plant (RCBD- ANOVA: F = 26; df = 16, 18; P < 0.001). No detectable increase in the population of plant parasitic nematodes was observed in healthy soil. The population of *Aphelenchoides sacchari* was significantly decreased in healthy plantation (F = 33.57; df = 2, 6; P < 0.001) as compared infected plants whereas the population of *A. sacchari* (F = 17; df = 2,6; P < 0.001) and *Hemicriconemoides mangiferae* (F = 23; df = 2,6; P < 0.001) were significantly decreased in healthy soil. The population levels of *Rotylenchulus reniformis* and *Helichotylenchus* were also considerably decreased (P < 0.001) in healthy plants. Significant differences were observed between healthy and infected plants on reduction of parasitic nematodes. Overall population of free-living or saprophytic nematodes was high and significantly increased in fertile soil due to which the vegetables and fruits were reproduce more and more vegetables and fruits. The population of *Acrobeles*, *Rhabditis*, and *Cephalobus*, soil nematodes were showed significantly increased in healthy and fertile soil (F = 9; df = 2,6; P < 0.0001); (F = 4; df = 2, 6; P < 0.001); (F = 8; df = 2,6; P < 0.00), respectively **Figure 2**. The overall results showed that the presence of abundant number of saprophytic nematodes indicates that the soil was filled with nitrogen, plant growth promoting rhizobacteria (PGPR) and fertility. Soil nematodes was highly active to fix the nitrogen fixing bacteria in soil. The population of *Acrobeles*, *Rhabditis*, and *Cephalobus*, nematodes species were comparatively more active nematodes for fixing and indicating the health of soil (**Figure 2**).

3.2 Nematode population in soil act as indicator

Rhabditis significantly (P < 0.001) showed the presence of higher level of nitrogen in soil whereas *Acrobeles* (A.) also significantly increased the nitrogen level (**Table 1** and **Figure 3**).

3.3 Plant root and shoot growth

Due to the presence of nematode associated bacteria the root, shoot length and number of forks was significantly (P < 0.01) increased shown in **Figures 1** and **4**.

S. No.	Plant and soil nematodes	Healthy plant	Infected plant
1	*Acrobeles*	75	15
2	*Aphelenchus*	12	45
4	*Aphelnchoides*	15	51
5	*Cephalobus*	59	25
6	*Cervidellus*	62	16
7	*Eucephalobus*	46	03
8	*Hoplolaimus*	19	28
9	*Heterocephalobus*	44	04
10	*Helichotylenchus*	21	55
11	*Meloidogyne*	00	10
12	*Paratylenchus*	09	37
13	*Pratylenchus*	01	18
14	*Plectus*	23	15
15	*Rhabditis*	82	20
16	*Trichodorous*	15	28
17	*Tylenchorhynchus*	24	65
18	*Xiphinema*	10	71

Table 1.
Percentage of plant parasitic and soil nematode population captured from healthy and infected plant.

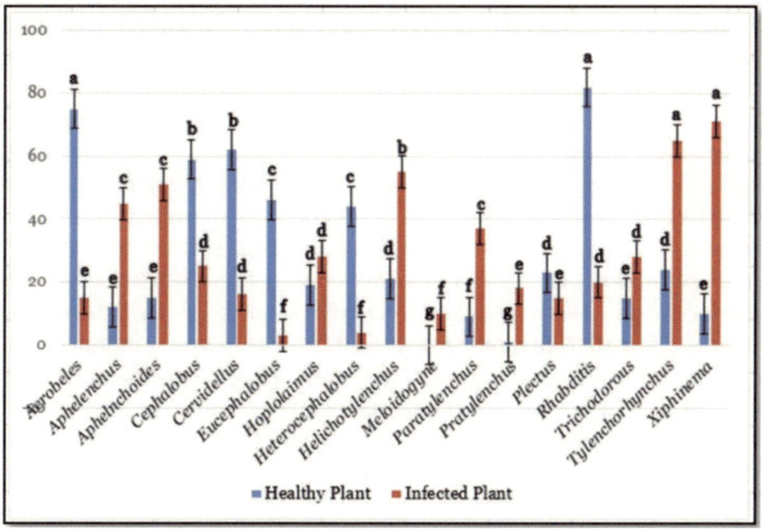

Figure 3.
Population density of plant and soil nematodes found from healthy and infected plants.

The indication of soil fertility and the suppression of plant parasitic nematodes also studies in this study. This observation was already reported by the previous researchers [33–37]. Our findings using saprophytic nematodes associated with bacteria also suppress the population of plant parasitic nematodes on *Cynodon dactylon* grass. Most plant parasitic nematode genera in our experiment was suppressed in

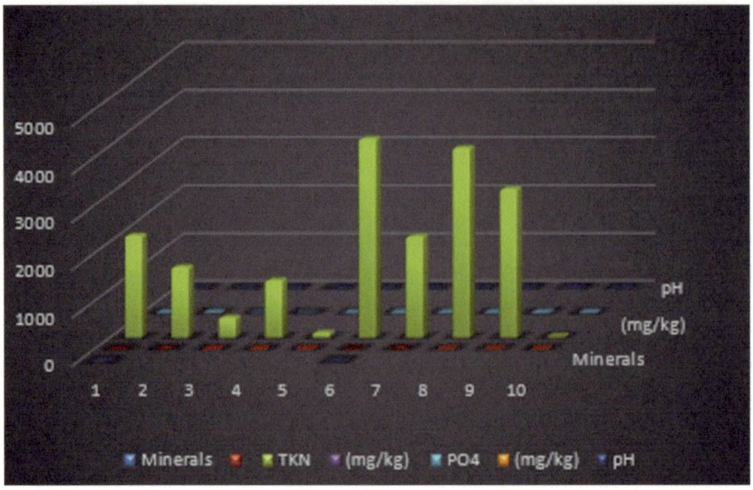

Figure 4.
Enhancement of nitrogen (TKN method) that showed by nematode as bio-indicator.

the abundant presence of free living bacterial feeding nematodes and also entomopathogenic nematodes. So it can evaluated the saprophytic nematodes should be as effective in suppressing the population of plant parasitic nematodes [35, 38].

For this purpose an experiment was conducted in which sterilized and non-sterilized soil by providing heat treatment were used. Four types of treatments with combination of control applied into soil (tomato pots) which has tomato seedlings. Treatments of nematodes culture obtained from the culture lab of NNRC, UoK. Nitrogen enhancement rates were calculated from soil. Different treatments differed in the amount of fixed nitrogen fixed in the soil, *Oscheius* A. treated pot show higher amount of nitrogen as compared to other treatments. The result of that study was conducted in green house condition where different factors were involved. The result of the experiment showed that *Acrobeles* significantly enhanced the soil fertility or nitrogen in soil [39].

The production of nematodes, which is bacterial feeding, nematodes directly associated to the rate of putrefaction of different organic modifications [40]. It is well known that saprophytic nematodes significantly increase soil nutrient absorption and bacterial population [41]. Inorganic nitrogen supports the plant growth initiates mostly from biotic activities in the soil. Thus indication about the richness, multiplicity and activities of different biotic fauna responsible for nitrogen mineralization is of vital position in the health of soil productivity. Soil organic materials could characterize the main source of inorganic nitrogen, even in the presence of fertilizer [42–44].

Nematodes, a diverse group of round worms, exist cosmopolitan in almost all biomes. Saprophytic free-living soil nematodes found as the part of agricultural fauna indicate an important role in the ecosystem. Usually 50 percent nematode fauna present in soil are saprophytic and the ratio reaches 80% at locations for high bacteriological population [40, 45–50] these are useful indicators of soil health because of their remarkable variety and their role in many functions of the soil food web. Many Scientists [51–53] have been studied and proved the evidence of the occurrence of saprophytic nematodes improved the nitrogen mineralization and later on stimulated plant growth experimented by different researchers [50, 54–57] and have indicating properties. They significantly enrolled in C mineralization and nutrient cycling, mainly by feeding on bacteria and fungi. Nematodes are the most abundant metazoans in soil.

Four of every five multi-cellular animals on the planets are nematodes. Normally twenty to fifty percent nematodes are bacterial feeders and the diversity of presence reaches 90–99% at locations of high bacterial activity [41, 46, 58]. These are the free-moving nematodes, not feeding on a particular plant but on the soil and bacteria. They are commonly associated with decaying root galls as probably they feed on decaying plant materials and increase soil fertility.

Different scientists of the world have been given the techniques to measuring the status of soil health by calculating the numbers of nematodes in different families in addition to their variety, they are beneficial indicators because of their population in response to fluctuations in moisture and temperature. Soil saprophytic nematodes preserved the level of plant -absorbable nitrogen in organic farming system. The process of recycling nutrients from organic to inorganic form is termed mineralization, Nematodes involved directly to nitrogen enhancement by their feeding interactions. For example nematodes ingest nitrogen in the form of proteins and other nitrogenous compounds and release extra amount of nitrogen as ammonia which is easily absorb for plant use. When nematodes graze on these microbes they give off CO2 and NH4 and increase soil fertility. Nematodes keep 1/6 of the nitrogen, they process and rest 5/6 is excreted to the soil for plant absorption. Classical management practices along with nematodes as bio-fertilizers are useful to increase soil conditions and crop productivity.

4. Conclusion

Results showed that *Acrobeles* and *Rhabditis* nematodes are significantly enhanced the soil fertility or nitrogen in soil. The increase in nematode numbers, especially bacterial feeding, especially bacterial feeding nematodes is directly associated to the rate of breakdown of different organic materials [40]. It is well recognized that soil nematodes significantly enhance nutrient in soil as well as bacterial populations [41]. The enhancing effect of bacterial-feeding nematodes on microbial population growth in soil microcosm has been reported by Mesfin *et al.*, [48], they found that all the treatments having nematodes and bacteria had higher bacterial densities than the treatments without nematodes. Our results supported this conclusion, suggesting that nematodes increased the bacterial densities, and populations of nematodes and bacteria rose simultaneously.

There is need to use the bacterial feeding nematodes as bio-fertilizer for production of healthy plants or crops. Based on the previous studies the practical use of nematodes seems to be more appropriate as they are effective to enhance nitrogen and carbon level in soil. Nematodes use as a bio-fertilizer gave benefits in agriculture to raise productivity. About thirty percent of the total inorganic nitrogen was mineralized in the form of soil organic matter that was for consumption soil microorganism [48, 59, 60]. Microphagous nematodes, establish an important group that effects on micro-organism activity and are important regulators of decomposition and nutrient release processes [41, 55]. Interactions between nematodes and microbes and have been studied under temperate soil conditions. Nematodes are present in diverse habitat they play the major role in the ecosystem advancement, soil properties, soil microbe's diversity, plant growth and crop production. The agrarian useful nematode fauna increase soil health with fixing the rhizobacteria, Nitrogen fixing cyanobacteria, plant beneficial bacteria and decomposition of microbes [1, 2, 61].

Author details

Tabassum Ara Khanum*, Nasir Mehmood and Nasira Khatoon
Department of Zoology, National Nematological Research Centre, University of Karachi, Karachi, Pakistan

*Address all correspondence to: tabassumak@uok.edu.pk

References

[1] Doran, J.W., Sarrantonio, M. & Liebig, M.A., (1996). Soil health and sustainability. In: Sparks, D.L. (Eds.), Advances in Agronomy. Academic Press, pp. 1-54.

[2] Elliot, E.T., (1997). Rationale for developing bioindicators of soil health. In: Pankhurst, C.E., Doube, B.M. & Gupta, V.V.S.R. (Eds.), Biological Indicators of Soil Health. CAB International, pp. 49-78.

[3] Doran, J.W., Safley, M., (1997). Defining and assessing soil health and sustainable productivity. In: Pankhurst, C.E., Doube, B.M., Gupta, V.V.S.R. (Eds.), Biological Indicators of Soil Health. CABI Publishing, Wallingford, UK, pp. 1-28.

[4] Pankhurst, C.E., Doube, B.M. and Gupta, V.V.S.R., 1997. Biological indicators of soil health: Synthesis. In: Pankhurst, C.E., Doube, B.M. & Gupta, V.V.S.R. (Eds.), *Biological Indicators of Soil Health*. CAB International, pp. 419-435.

[5] Haitova, D. and Bileva T. (2011). Influence of different fertilizer types of zucchini (*Cucurbita pepo*) on the structure of nematode communites. "Comm. in Agric. and Appl. Biol. Scien.", Ghent University, vol. 76 (3): 341 – 345.

[6] Deepali and Gangwar K.K. (2010). Biofertilizers: An ecofriendly way to replace chemical fertilizers. http://www.krishisewa.com/cms/articles/2010/biofert.html.

[7] Wang, K. and Hooks C. (2011). Chapter 4: Managing soil health and soil health bioindicators through the use of cover crops and other sustainable practices. In: G.E. Brust (ed.) *MD Organic Vegetable Growers*.

[8] Gaur, V. (2010). Biofertilizer-Necessity for Sustainability. J. Adv. Dev., 1, 7-8.

[9] Vaids, S.; Shah, A.A.; Ahmad, R.; Hussain, A. (2014). Diversity of soil inhabiting nematodes in Dera Ki Gali forest of Poonch district, Jammu and Kashmir, India. International Journal of Nematology, v.24: n.1, p.97-102.

[10] Doran J.W. and Zeiss, M.R. (2000). Soil health and sustainability: managing the biotic component of soil quality. Applied Soil Ecology 15: 3-11.

[11] Neher, D.A., Wu, J.H., Barbercheck, M.E. and Anas, O., (2005). Ecosystem type affects interpretation of soil nematode community measures. Applied Soil Ecology 30: 47-64.

[12] Neher, D.A., (2010). Ecology of plant and free-living nematodes in natural and agricultural soil. Annual Review of Phytopathology 48: 371-394.

[13] Bileva T., Zh Arnaudova,. (2011). Mapping of nematode distribution and assessment of its ecological status using GIS techniques in Plovdiv region, Bulgaria. "Comm. in *Agric. and Appl. Biol. Scien.*", Ghent University, vol. 76 (3) 347-353.

[14] Pattison, A.B., Moody, P.W., Badcock, K.A., Smith, L.J., Armour, J.A., Rasiah, V., Cobon, J.A., Gulino, L.M. & Mayer, R., (2008). Development of key soil health indicators for the Australian banana industry. Applied Soil Ecology 40: 155-164.

[15] Yeates, G.W.; Bongers, T. (1999). Nematode diversity in agroecosystems. *Agriculture Ecosystems Environment*, v.74: p.113-135.

[16] Borris R. (2011). Use of plant-associated *Bacillus* strains as biofertilizers and biocontrol agents in agriculture. In: Maheshwari D.K. (ed) *Bacteria in agrobiology: Plant growth responses*. Springer Verlag Berlin Heidelberg.

[17] Boltovskoy, E. (1965). Los Foraminıferos Recientes. Editorial Universitaria de Buenos Aires (EUDEBA), Buenos Aires.

[18] Resig, J.M. (1960). Foraminiferal ecology around ocean outfalls off southern California. In: Disposal in the Marine Environment. Person, E. (Ed.), pp. 104-121, Pergamon Press, London.

[19] Watkins, J.G. (1961). Foraminiferal ecology around the Orange County, California, ocean sewer outfall. *Micropaleontology*, Vol. 7: pp. 199-206.

[20] Alve, E. (1995). Benthic foraminifera response to estuarine pollution: a review. Journal of Foraminiferal Research, Vol. 25: pp. 190-203.

[21] Boltovskoy, E., Scott, D.B., & Medioli, F.S. (1991). Morphological variations of benthic foraminiferal test in response to changes in ecological parameters: a review. Journal of Paleontology, Vol. 65: pp. 175-185.

[22] Frontalini, F., & Coccioni, R. (2011). Benthic foraminifera as bioindicators of pollution: A review of Italian research over the last three decades. Revue de Micropaléontologie, Vol. 54: pp. 115-127.

[23] Murray, J.W., & Alve, E. (2002). Benthic foraminifera as indicators of environmental change: marginal-marine, shelf and upper-slope environments. In: Quaternary Environmental Micropalaeontology, Haslett, S.K. (Ed.), pp. 59-90, Edward Arnold (Publishers) Limited, London.

[24] Nigam, R., Saraswat, R., & Panchang, R. (2006). Application of foraminifers in ecotoxicology: retrospect, perspect and prospect. Environmental International, Vol. 32: pp. 273-283.

[25] Yanko, V., Kronfeld, J., & Flexer, A. (1994). Response of benthic foraminifera to various pollution sources: implications for pollution monitoring. Journal of Foraminiferal Research, Vol. 24: pp. 1-17.

[26] Soko, M.I. and Gyedu-Ababio, T.K. (2019) Free-Living Nematodes as Pollution Indicator in Incomati River Estuary, Mozambique. Open Journal of Ecology: 9, 117-133. https://doi.org/10.4236/oje.2019.95010

[27] Chander, K. & Brookes, P.C., (1993). Residual effects of zinc, copper, and nickel in sewage sludge on microbial biomass in a sandy loam. Soil Biology and Biochemistry 25: 1231-1239.

[28] Cobb, N.A. (1918). Estimating the nema population of soil. Agric. Tech. Circ. U. S. Dept. Agric., 1: 48 pp.

[29] Baermann, G. (1917). Enine method zur Auffindung von Anklyostomum (Nematoden) larven in *Erproben.* Geneesk Tij dschr. Ned. Ind., 57: 131 –137.

[30] Dick, R.P., (1997). Soil enzyme activities as integrative indicators of soil health. In: Pankhurst, C.E., Doube, B.M. & Gupta, V.V.S.R. (Eds.), Biological Indicators of Soil Health. CAB International, pp. 121-156.

[31] Hornby, D. & Bateman, G.L., (1997). Potential use of plant root pathogens as bioindicators of soil health. In: Pankhurst, C.E., Doube, B.M. & Gupta, V.V.S.R. (Eds.*), Biological Indicators of Soil Health*. CAB International, pp. 179-200.

[32] Kjeldahl, J. (1883). Neue Method zur Bestimmung des Stickstoffs in organism Korpern (New method for the determination of Nitrogen in organic substances), Zeitschriftfuranalytishe Chemie 22: 366-383.

[33] Bird, A. and Bird, J. (1986). Observations on the use of insect parasitic nematodes as a means of

biological control of root-knot nematodes. Int. J. Nematol., 26: 127- 137.

[34] Grewal P.S. and Georges, R. (1998). Entomopathogenic nematodes. 271-299 pp. In: Methods in Biotechnology Biopsticides: Use and Delivery Vol.5. (Eds). F.R. Hall and J.J. Menn, Humana Press, Totowa, N.J.

[35] Ishibashi, N. & Kondo, E. (1986). *Steinernema feltiae* (DD-136) and *S. glesseri*: persistence in soil and bark compost and their influence on native nematodes. J. Nematol., 9: 404-412.

[36] Lewis, E.E. and Grewal, P.S. (2005). Effects of entomopathogenic nematodes on plant parasitic nematodes. 349-361 pp. In:Nematodes as Biocontrol Agents (Eds.) P.S. Grewal, R.U. Ehlers and D. Shapiro-Ilan, CABI Publishing, Wallingford, UK.

[37] Yousef, M.M., & M.F.M. Essa 2014. Biofertilizers and their role in management of plant parasitic nematodes. A review. *E3 J. Biotech. and Pharmaceut. Res.*, 5(1). pp.001-006.

[38] Janvier, C., Villeneuve, F., Alabouvetter, C., Edel-Hermann, V., Mateille, T. & Steinberg, C., (2007). Soil health through soil disease suppression: Which strategy from descriptors to indicators? Soil Biology & Biochemistry 39: 1-23

[39] Shahina, F., and K.A. Tabassum (2009). Suppression of plant parasitic nematodes in bermuda grass using live or dead entomopathogenic nematodes *Steinernema pakistanense Pak*. J. Nematol.,27: 167-178.

[40] Griffiths B.S., Ritz, K., Wheatley, R.E. (1994). Nematodes as indicators of enhanced microbiological activity in a Scottish organic farming system. *Soil Use and Manag*.,10: 20-24.

[41] Griffiths, B.S. (1994). Microbial feeding nematodes and protozoa in soil: their effects on microbial activity and nitrogen mineralization in decomposition hotspots and the rhizosphere. Pl and Soi., 164: 25-33.

[42] Chotte, J.L., Feller, C., Hetier, J.M. and Mariotti, A. (1998). The fate of fertilizer nitrogen under maize cropped after different land use histories. Field studies in the volcanic Lesser Antilles with 15N-urea. Tropical Agriculture 75 : 330-336.

[43] Guiraud, G. (1984). *Contribution* du *marquage isotopique* h l'*Evaluation* des *transferts d'azote entre* les *compartiments organiques et mineraux dans les systmes sol-plante*. Thse de Doctorat d'Etat, Universite Pierre et Marie Curie, Paris VI, France, 335 p.

[44] Song, M., X. Li, S. Jing, L. Lei, J. Wang, and S. Wan. 2016. Responses of soil nematodes to water and nitrogen additions in an old-field grassland. *Applied Soil Ecology*, 102: 53-60.

[45] Hu, C., Xia, X.G., Han, X.M., Chen, Y.F., Qiao, Y., Liu, D. H., Li, S.L. (2018). Soil nematode abundances were increased by an incremental nutrient input in a paddy-upland rotation system. Helminthologia, 55: 4: 322 - 333.

[46] Li, H.X., Kazuyuki, I., and Johji, M., (2001). Effects of temperature on population growth and N mineralization of soil bacteria and a bacterial-feeding nematode. Micro and Environ., 16: 141-146.

[47] Maria Balsamo, Federica Semprucci, Fabrizio Frontalini and Rodolfo Coccioni (March 2nd 2012). Meiofauna as a Tool for Marine Ecosystem Biomonitoring, Marine Ecosystems, Antonio Cruzado, IntechOpen, DOI: 10.5772/34423.

[48] Mesfin T. Gebremikael, Hanne S., David B., Wim B. and Stefaan D.N. (2018). Nematodes enhance plant growth and nutrient uptake under C

and N-rich conditions. *Scientific Reports* | 6:32862 | DOI: 10.1038/srep32862

[49] Palwasha R., Rashid N., Tatheer A.N., Arshid P., and Usman I. (2018). Bacterial feeder Neher, D.A., Wu, J.H., Barbercheck, M.E. & Anas, O., 2005. Ecosystem type affects interpretation of soil nematode community measures. Applied Soil Ecology, 30: 47-64.

[50] Xiao Hai-Feng, L.I.,Gen, L.I., Da-Ming, Hufeng and L.I., Hui-Xin 2014. Effect of Different Bacterial-Feeding Nematode Species on Soil Bacterial Numbers, Activity, and Community Composition. *Pedosphere* 24(1): 116-124.

[51] Anderson, R.V., Cole, C.V. and Coleman, D.C. (1980). Quantities of plant nutrients in the microbial biomass of selected soils. Soil Sciences 130: 211-216.

[52] Ferris, H., Venette, R.C., van der Meulen, H.R. and Lau, S.S. (1998). Nitrogen mineralization by bacterial feeding nematodes: verification and measurement. Plant and Soil 203: 159-171.

[53] Hu, F., Li, H.X., Xie, L.Q. and Wu, S.M. (1999). Interaction of bacterivorous nematodes and bacteria and their effects on mineralization-immobilization of nitrogen and phosph orus.*ActaEcologicaSinica* 19: 914-920 (in Chinese, with English abstract).

[54] Bonkwski, M., Griffiths, B.S. and Scrimgeour, C. (2000). Substrate heterogeneity and microfauna in soil organic hotspots as determinants of nitrogen capture and growth of rye-grass. Applied Soil Ecology 14: 37-53.

[55] Ingham, R.E., Trofymow, J.A., Ingham, E.R. and Coleman, D.C. (1985). Interaction of bacteria, fungi, and their nematode grazers: effects on nutrient cycling and plant growth. Ecological Monographs 55: 119-140.

[56] Knox, O.G.G., Killham, K., Mullins, C.E. and Wilson, M.J. (2003). Nematode enhanced microbial colonization of the wheat rhizosphere. FEMS Microbial Lett., 225, 227-233.

[57] Li, H.X. and Hu, F. (2001). Effect of bacterial feeding nematode inoculation on wheat growth and N and P uptake. Pedosphere 11: 57-62.

[58] Gabriela, S. M. and Gilmar, F. (2017). Biodiversity of nematodes biological indicators of soil quality in the agroecosystems. Arq. Inst. Biol., 84: 1-8. DOI: 10.1590/1808-1657000142015

[59] Janina Schenka, Sebastian Hössa, Marvin Brinkec , Nils Kleinböltingd , Henrike Brüchner-Hüttemanna , Walter Traunspurgera (2020) Nematodes as bioindicators of polluted sediments using metabarcoding and microscopic taxonomy *Environment International* 143 105922 https://doi.org/10.1016/j.envint.2020.105922

[60] Tabassum Ara Khanum and Nasir Mehmood. (2021) "Bacterial Feeding Nematodes Use for Nitrogen Mineralization and Plant Production". Acta Scientific Pharmaceutical Sciences 5.5 : 02-06.

[61] Tatyana Bileva, Vera Stefanova, and Dimka Haytova (2014) Assessment of Nematodes as Bioindicators of Soil Health in Agroecosystems. Turkish Journal of Agricultural and Natural Sciences Special Issue: 1

Section 5

Entomopathogenic and Marine Nematodes

Chapter 11

Entomopathogenic Nematodes: Biological Model of Studies with Anthelmintics

Oscar Barrón-Bravo, Ismael Montiel-Maya, Ana Cruz-Avalos, Fidel Avila-Ramos, Jaime Molina Ochoa and César Angel-Sahagún

Abstract

Anthelmintics used in animals to combat parasitic infections are mainly excreted in manure and cause negative effects on the environment and decomposers. Nematodes are associated with the rhizosphere; some are gastrointestinal parasites of animals, and others regulate insects and other arthropods (entomopathogenic nematodes) and are considered beneficial. The habitat and the similarities that exist among them give the opportunity to use nematodes as a biological model. The availability of target organisms is not always feasible; therefore, experimental studies with models similar to those of the target organisms are a possibility. In veterinary clinics, the study of drug susceptibility is a fundamental tool to monitor the development of resistance. To conserve the biodiversity of the environment, it is necessary to make adequate use of anthelmintics, avoid resistance to these pesticides and prevent the used products from damaging populations of beneficial organisms.

Keywords: Rhizosphere, parasites, intestinal, entomopathogens, anthelmintics, resistance

1. Introduction

Nematodes in nature show astonishing biodiversity since only approximately 30,000 species have been described, but it is estimated that there are a million or more species of nematodes worldwide. Nematodes generally have a long, narrow and thread-like body ('nema' from the Greek 'thread'), not segmented like earthworms. Its body is basically tubular, and the intestine and gonad are surrounded by the wall of the body with its dorsal and ventral longitudinal muscles, epidermis and cuticle. Between the inner and outer tubes, there is a pressurised cavity filled with fluid that acts as a hydrostatic skeleton, all of which allows the nematodes to move in sinusoidal waves. The morphological diversity in this group is restricted and much lower than that of other arthropods or vertebrates. All nematodes go through three or four larval stages, and at the end of each stage, a new cuticle is synthesised, and the previous cuticle is moulted. The nematode species are very diverse, and the most obvious differences are observed in size, which varies from fractions of a millimetre to several metres, cuticular decorations and especially feeding structures.

The mouth of nematodes can be a simple tube, or it can be decorated with a perforating stylet (in plant parasites and fungal feeders) or with teeth that can cut, tear or bite in predatory species such as *Mononchus* and in some intestinal parasites such as *Strongylus* [1].

Nematodes are also the most numerous group of parasites of animals and humans and are widely distributed in a variety of habitats. Some are free-living, while others are in some part of their life cycle parasites of plants and vertebrates or invertebrates [2]. Parasitic nematodes of animals have great economic importance; due to the high frequency with which they occur, they are generally chronic and most interfere with good growth. They can be located in most organs, mainly in the digestive tract, have a direct or indirect life cycle, and some are zoonotic [3].

2. Anatomy

The cuticle is a noncellular, flexible and elastic structure that generally has externally arranged rings; however, since the cuticle is not visible, it has a smooth, shiny appearance and is secreted by the layer of cells that are immediately below, that is, the hypodermis (**Figure 1**). The cuticle is formed by several layers whose number and thickness vary according to the species in question and is composed of proteins such as albumin and glycoproteins [4].

The hypodermis is a thin layer of four tubular thickenings, called the dorsal cord, two lateral and one ventral. It contains cells that secrete the layers of the cuticle. The muscular system is composed of two types of muscles, specialised and nonspecialized or somatic, which occupy a position close to the hypodermis of the areas between the cords, forming a single layer of cells, which has an important role in body movements [5].

The specialised muscles are found in several positions and have important functions, such as the oesophageal muscles in the wall of the oesophagus, the intestinal muscles in the wall of the intestine, the dilator and compressor muscles of the

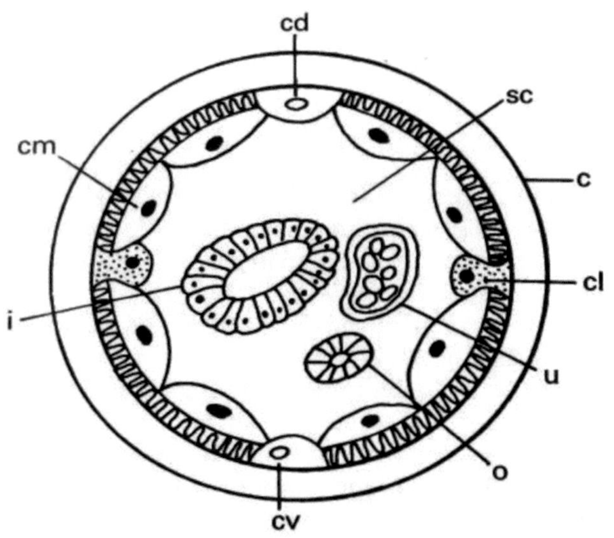

Figure 1.
Nematode cross section (c) cuticle, (cd) dorsal field, (cl) lateral field, (cm) muscle cell, (cv) ventral field, (i) intestine, (o) ovary, (u) uterus.

anus, the copulatory muscles, those of the copulatory bursa, of the spicules, of the gubernaculum and vulva [4].

Cordero del Campillo *et al.* [6] described the anatomy of nematodes in detail regarding the digestive system of nematodes, indicating that it is elongated and has a sac-like shape; it is composed of different organs (**Figure 2**). The mouth is located in the subdorsal or ventral apex, and the primitive model is made up of six lips with two papillae each distributed in two circles (internal and middle) and a third or external circle of four papillae and two lateral amphids, although there are extensive variations in morphology and position. The oral cavity or orifice is a dilation in which hooks or teeth are found both in the cuticular walls in the oral capsule or in the bottom of the cavity. The oesophagus is a radiated muscular organ covered by a thick cuticle, and the muscles that occupy the oesophageal lumen contain three glands, which produce enzymes for its digestive function (**Figure 3**). The intestine

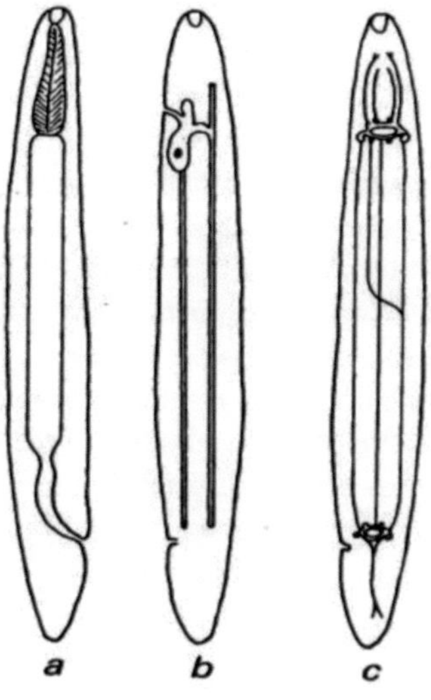

Figure 2.
Schematization of (a) digestive system, (b) excretory system, (c) nervous system.

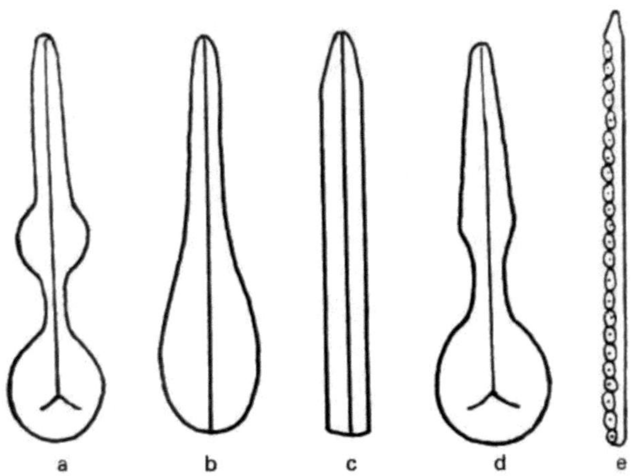

Figure 3.
Different types of nematode oesophagus; (a) rhabditoid, (b) strongyloid, (c) filariform, (d) oxyuride, (e) trichuroid.

is cylindrical in shape and is composed of a basal lamina and a single cell epithelium with a nonmuscular wall covered by microvilli (**Figure 2**). Finally, the rectum is found as an invagination of the cuticular lining that in males gives rise to the cloaca; in some species, it also has glands. In addition to the functions of absorption and defecation of the rectum, it fulfils other functions as an organ of secretion.

The excretory system contains two unbranched lateral tubes that are part of the lateral cords of the hypodermis. The excretory duct comprises the transverse canal to the excretory pore, which generally encompasses the cephalic and cervical region of the nematode.

The nervous system of nematodes is formed by the circumoesophageal ring, which contains ganglia and surrounds the oesophagus, with numerous longitudinal nerves (**Figure 2**). On the dorsal and ventral lines of the body of the nematodes are two of the six short nerves that pass through the anterior part of the body, and another six long nerves pass to the posterior end of the body, likewise a shorter nerve in the ventral line, and the longest in the dorsal line and two small ones in each lateral line. The main nerves emit nerve fibres that can become entangled throughout the body and rejoin these nerves or follow other routes. In general, the sensory organs of parasitic nematodes are less developed than those of nonparasitic nematodes since parasitic life degenerates and atrophies the sensory organs. For these, if we can find the oral papillae, a pair of cervical papillae, and amphidal papillae located in the anterior end, in the posterior end of the body of the male, we can find the genital papillae and other called phasmids in the posterior end of males and females, we can also find an anal ganglion.

The reproductive organs of nematodes are filiform, whitish, long and spirally wound, and their apical end is blind. They continue with long tubes of similar morphology that lead them to the genital cells to the outside. Both the ovaries and the testicles begin as thin threads and are transformed into a central cord, and the surrounding genital cells are later transformed into genital ducts. The male has only one testicle and a vas deferens through which sperm discharge, a seminal vesicle where sperm are stored, and an ejaculatory duct that ends in the cloaca. The testes of most nematodes are of the telegonic type; we also found spermatogonia that extend from the distal portion of the tube and complete along the walls of the central rachis. The spicules are the copulatory organs that are generally found in nematodes, which are elongated and filiform organs of varied dimensions. The spicules are formed in the dorsal sac of the cloaca and are formed by cuticular material. The retractor and ejector muscles are responsible for supporting the spicules, and the gubernaculum is the accessory organ on which the spicules slide and are oriented into and out of the cloaca. Some species also have a caudal pouch, which helps the male attach himself onto the female, and those that do not have it have other cuticular structures, grooves and rough areas.

The female genital tract is formed by the ovary that goes through a maturation process where the oogonia begin in the germinal zone and end in the maturation zone. The oviduct is a short and narrow tube through which the oocytes pass, containing epithelial cells at its base. The seminal receptacle is widening at the beginning of the uterus where sperm are stored. The uterus has an epithelial layer, a basal lamina and muscular annular cells. The vagina is covered in its proximal part by a cuticle, and the vaginal opening is in the ventral medial line of the helminth. The reproductive system of females can be monodelphic, didelphic or polydelphic.

The eggs of the nematodes have a more or less oval shape, and depending on the species, they also vary in size and content. Generally, we find three layers: the lipid layer, the chitinous layer and the external or vitelline layer.

3. Life cycle

Generally, parasitic nematodes are sexually reproduced; males produce sperm, and females produce ovules, which are fertilised after copulation. Embryonic development includes the stages of morula, blastula, gastrula, and tadpole where the embryo acquires its shape. The life cycle includes an egg stage, larval stages (three or four) and an adult stage. Between each larval stage, there is a moulting or change in the cuticle, which can be rigid or elastic and allow growth. Through enzymatic action, each larval stage is released from its envelope to reach the next stage, which may be preceded by lethargy. The life cycles may or may not have one or more intermediate hosts, and the eggs or larvae produced in the definitive host are not infectious, except for rare exceptions. Larval development need to reach the infective stage. In the direct cycles, this development occurs in wet soil, prairies or water. In the indirect cycles, the development up to the infective stage occurs in the intermediate host. In the direct cycle, infestation is usually orally through the ingestion of eggs or larvae; and in the indirect cycle, it can be orally through the ingestion of the intermediate host or arthropod bites [5].

Larval migration to reach the site where they reach sexual maturity can occur through the digestive, hepato-cardio-pulmonary or lymphatic-cardio-pulmonary tracts. The process in which a developmental stage reaches another host includes a complex system of relationships between the animal population and the environment, which vary in time and space. The influence of the environment is important with factors such as temperature, humidity, luminosity, winds, rainfall, types of soil, types of vegetation and seasonal variation. Direct sun rays and dehydration rapidly destroy the larval stages, and the temperature has a range in which the conditions are optimal; outside this range, the physiological process stops and can be destroyed [7].

4. Entomopathogenic nematodes

The entomopathogenic nematodes of the families Steinernematidae and Heterorhabditidae are obligate parasites of arthropods, which live in the soil and are ubiquitous and are used commercially to suppress insect pests that live in the soil in agricultural fields [8]. Their use is incipient in the veterinary field.

Nematodes of the Steinernematidae family are characterised by their mutualistic association with bacteria of the genus *Xenorhabdus*. This family is currently comprised by two genera, Steinernema, with more than 70 species, and Neosteinernema, with a single species, *N. longicurvicauda* [1].

The family Heterorhabditidae consists of one genus, *Heterorhabditis*, with *H. bacteriophora* as the model and 17 other species described. These have a life cycle similar to that of nematodes in general, but the adults that result from infectious juveniles (IJs) are hermaphrodites. The eggs laid by the hermaphrodites produce juveniles that become males and females or IJ. Males and females mate and produce eggs that develop in IJ [1].

Natural entomopathogenic nematodes can suppress insects in a wide variety of ecosystems. Insects at any stage of life that come into contact with infested soil are potentially susceptible to infection, and persistent populations of entomopathogenic nematodes in agricultural systems can provide valuable assistance to producers by reducing the costs associated with the management of insect pests. Knowing the environments in which entomopathogenic nematodes persist successfully helps to conserve the natural populations of these insect pathogens that are potentially valuable for agricultural production [8].

4.1 Isolation of entomopathogenic nematodes

Because entomopathogenic nematodes are found in the rhizosphere, moist soil must be sampled for their isolation; it must be sieved to keep it free of organic matter and placed in containers adding larvae of the last instar of greater wax moth *Galleria mellonella* (Linnaeus), which has the objective of being a trap insect to complete the isolation. *G. mellonella* larvae are susceptible to infection by entomopathogenic nematode species [4].

The container with the soil and the larvae of *G. mellonella* is covered and inverted and incubated at 25 ± 1°C for seven days. After this time, the larvae and/or pupae are recovered and placed in Petri dishes with a double layer of wet filter paper to maintain a high relative humidity and favour the development of the infection [9, 10].

Once nematode infection is observed on the dead larvae, it is transferred to a White trap, which helps to separate infectious juvenile nematodes [11]. With the result of this isolation, they reproduce again in other larvae of *G. mellonella* to make their identification and later use them for the necessary assessments.

5. Anthelmintics

Anthelmintics are drugs used in the treatment against parasitic diseases, and they continue to be the cornerstone of parasite control programmes in animals; however, their irrational use has led parasites to develop resistance [12].

Since the parasites are grouped into three categories, Nematoda, Cestoda and Trematoda, there are also three categories or groups of drugs that are available for their treatment. Nematocidal drugs against intestinal worms, hookworms, Ascaris and Strongyloides include piperazine, mebendazole, thiabendazole, pyrantel, ivermectin and diethylcarbamazine, among others. Antitrematodal drugs include praziquantel, bithionol sulfoxide, oxamniquine, and metrifonate. The third group of antihelminths are anticestodal drugs, such as niclosamide, which are applied against *Taenia, Echinococcus* and *Diphyllobothrium*. Levamisole is often prescribed as an antiparasitic drug against nematodes such as Ascaris and tricostrongyloid species [13].

The most commonly used anthelmintics, their mode of action and mechanism of excretion to the environment are listed in **Table 1**.

5.1 Mechanisms of excretion of some chemicals used as anthelmintics

Because there is a diversity of anthelmintics, only some of the currently most used will be described as an example of the studies that have been performed for each of the active ingredients available.

There are several pharmacokinetic cycles for anthelmintics in animals, from which it is derived that excretion varies; for example, thiabendazole follows an enterohepatic cycle, the amount that is absorbed is rapidly metabolised in the liver by hydroxylation, and its main metabolite is 5-hydroxythiabendazole, which is also metabolised by glucuronidation and sulfate formation. After 8 h of its administration, 90% of the drug is eliminated as a metabolite through the urine and 5% in the faeces. Five days after the last dosage, it was completely eliminated from the body [13].

The fenbendazole that is absorbed is metabolised (and vice versa) and converted into oxfendazole (active compound), fenbendazole sulfone, fenbendazole-2-aminosulfone and other minor metabolites. The drug that is not absorbed (most of it) is eliminated in faeces and the rest in urine and milk, where 0.3% of the applied dose is detected. In sheep, cattle, and pigs, 44 to 50% of the fenbendazole dose is excreted unchanged in the faeces and less than 1% in urine [32, 33].

Chemical group	Drug	Target/mode of action	Main mechanism of excretion
Aminoacetonitriles	Monepantel*	nAChR allosteric modulator	Urine and feces [14, 15]
Benzimidazoles	Albendazole* Cambendazole Fenbendazole Mebendazole* Oxfendazole Oxibendazole Parbendazole Thiabendazole* Triclabendazole	β-Tubulin inhibitor Fumarate reduct ase inhibitor β-Tubulin inhibitor	Urine and feces [14, 15]
Benzimidazoles (pro-)	Febantel Netobimin Thiophanate	β-Tubulin inhibitor	Urine and feces [16, 17]
Cyclooctadepsipeptides	Emodepside	LAT-1/SLO-1 inhibitor	Urine and feces [18]
Imidazothiazoles	Tetramisole* Levamisole*	L-subtype nAChR agonist	Urine [19, 20]
Macrocyclic lactones	Abamectin* Doramectin Ivermectin* Moxidectin* Nemadectin	Glutamate- and GABA-gated chloride channels receptor agonist	Urine and feces [21–23]
Organophosphates	Dichlorvos Haloxon Trichlorfon	Acetylcholinesterase inhibitor	Urine [24]
Pyrazinoisoquinolines	Phenothiazine Piperazine Praziquantel	GABA receptor agonist Depolarization of the tegument. Rapid levels of Ca2+ in the sarcoplasmic reticulum	Urine [25, 26]
Salicylanilides	Closantel Niclosamide Oxyclozanide Rafoxanide	Uncoupler of the oxidative phosphorylation	Urine and feces [23, 27]
Spiroindoles	Derquantel*	nAChR antagonist	Urine [28]
Substituted phenols	Bithionol Nitroscanate Nitroxynil	Uncoupler of the oxidative phosphorylation	Urine and feces [18]
Tetrahydro-pyrimidines	Morantel* Pyrantel pamoate* Pyrantel tartrate	L-subtype nAChR agonist	Urine and feces [23, 29, 30]

Note: The compounds which are marked with an asterisk are also used in humans.
Abbreviations: GABA, γ-aminobutyric acid; LAT-1, latrophilin-1; nAChR, nicotinic acetylcholine receptor; SLO-1, slowpoke potassium channel type 1.
Table adapted from Sepúlveda-Crespo et al. [31].

Table 1.
Anthelmintics most commonly used for treatment in humans and veterinary medicine, mode of action and mechanisms of excretion to the environment.

For closantel, its elimination is up to 75% in faeces and to a lesser amount in urine 0.57, 50% of the administered dose is eliminated in 50 to 85 hours, excreting up to 90% of the dose unchanged [27, 34].

Regarding the complete cycle of pharmacokinetics in animals, ivermectin, for example, is a commonly used drug that has been developed by laboratories for its application by different routes (subcutaneous, oral and topical). The oral route shows lower bioavailability, but its values in plasma can last from seven to 14 days, so in low doses (10–40 μg/kg/day), it can be very effective for the control of infestations by parasites (the intramuscular route is not recommended). The absorption processes show differences according to the routes of application and the species treated. Some oily preparations applied subcutaneously reach therapeutic concentrations of 80 to 90 days with a half-life of 36 hours [35]. It has a high volume of distribution with slight variations between species. Because it is a natural lipophilic substance, it is widely distributed in all tissues and tends to accumulate in adipose tissue. The highest concentrations are found in the liver, bile and skin, while the lowest concentrations are in the brain. It is poorly metabolised in the body; therefore, a large part of the dose is excreted unchanged [22].

It has been detected that the gastric content has the highest concentration of the drug. On the other hand, it is concentrated in large amounts in the mucus and intestinal content, so it is feasible to recover a large amount in the faeces, regardless of its route of administration. Ivermectin metabolism is carried out by hydroxylation processes in the rumen, stomach or intestine, regardless of the route of administration. Its metabolites are 24-hydroxymethyl-H2B1a, and 24-hydroxymethyl-H2B1b is eliminated by bile, so large amounts are detected in faeces, although it is also excreted in urine (2%) and in milk. Faecal excretion represents 90% of the total administered dose and in cattle up to 98% or more [35].

In horses, unlike ruminants, the absorption process is faster after oral administration than with subcutaneous administration, and although the injection results in a greater bioavailability, the oral route is preferred, since parenteral administration can produce local swelling and other adverse reactions. Plasma concentrations are higher and are reached more quickly in horses than in sheep, probably because the rumen delays absorption in ruminants. However, the half-lives of sheep in subcutaneous and oral application were 3.7 and 2.8 days, respectively, similar to those of sheep. In horses, the mean residence time is also longer after oral administration (4.2 days) and subcutaneous administration (3 days) and longer in donkeys (6.5 days), with a half-life of 7.4 days. In horses treated subcutaneously, most of the dose (90%) was excreted faecally in 4 days. The higher concentrations found in equine faeces compared to cattle faeces have been attributed to a lower production of more concentrated faeces [22].

Once in the environment, ivermectin can be rapidly degraded when exposed to sunlight. This photodegradation occurs in the presence of ultraviolet light and can occur between 0.5 and 23 days, a period in which it can affect living beings that have contact with the drug, such as earthworms, beetles, insects, fish and even humans [36]. If the meat or by-products of treated animals are consumed by humans, it usually constitutes a public health problem. The residual effect of the drug can be 10 to 12 weeks, and this is considered ideal for the control of ectoparasites, such as fleas, ticks or flies [35].

6. Tests to determine the resistance of parasitic nematodes of animals

6.1 *In vitro*

6.1.1 *Larval motility test*

The test was performed in a flat-bottomed cell culture plate; to facilitate the procedure, a 24-well plate was used. Nematodes are applied, and counts are performed in each of the wells. Finally, the treatments are inoculated to avoid mortality due to inadequate management [37]. The plate was incubated at 25 ± 1°C for 24 hours in complete darkness, after which a second nematode count was performed to determine the living and dead individuals in each well [38].

6.1.2 Larval migration inhibition test

For the larval migration inhibition test, a migration system is used that allows the physical separation of IJ with motility of the immobile ones through a 25 μm polypropylene mesh filter. The diameter of the pores allows active larvae, but not dead larvae, to pass through the mesh. In a new plate, the mesh filter is placed in each well, and the total volume of each well is transferred from the plate used in the motility test. The samples are incubated for 24 hours at 25 ± 1°C in complete darkness, and then the live and dead nematodes were counted [39].

6.2 *In situ*

6.2.1 Faecal egg count reduction

The egg counts of parasitic nematodes present in the excrement are considered the main test for parasite control because it has been shown that animals maintain relatively consistent levels of egg excretion over time.

In this technique, a suspension of faecal material is dispersed in a solution of higher density than the eggs of parasites (solution with common salt, 33% zinc sulfate, 35% magnesium sulfate, saturated sugar solution, sodium nitrate, etc.). The difference in specific gravity causes the eggs to rise to the surface or all float to the same level. The solution mixed with the excrement is allowed to settle, and most of the faecal particles will fall towards the bottom since their density is greater than that of the solution. This step is important for some parasitological diagnostic techniques but not for this test. Therefore, the procedure should be completed in the shortest possible time or regularly homogenise the mixture.

To achieve the procedure and determine the reduction of eggs per gram of excreted faeces, it is necessary to use the parasitological technique that uses the device called McMaster chamber where the number of eggs in a given amount of liquid (0.15 mL) is verified and then procedures to estimate the amount of parasitic nematode eggs per gram of excrement used initially are performed, this activity must be done before and after treatments with anthelmintics, and subsequently calculate the percentage of reduction of the egg count with the following formula:

$$FECR = \left(\frac{(EPG\ pretreatment - EPG\ postreatment)}{EPG\ pretreatment} \right) x\,100 \quad (1)$$

This technique is commonly used in the initial tests when populations of chemically resistant nematodes are suspected.

7. Biological models used for the assessment of anthelmintics

There are many limitations to conducting experimental studies on parasitic nematodes to assess the anthelmintic potential of a new product or drug, both for in vitro studies and in vivo studies. The difficulties, among others, are the difficulty of evaluating their activity in adult stages of parasitic nematodes kept outside the host, the cost of infection and sacrifice of experimental donor animals, and the impossibility of obtaining large quantities of the stages of the nematodes under study., the cost of maintaining the hosts, labour, the total time of handling the animals, and the approval of different ethics committees, among others not listed [40–42].

Although there are problems because in vitro studies with biological models different from parasitic nematodes sometimes do not offer reliable results, the use of entomopathogenic nematodes can help to standardise studies or preliminary tests that allow establishing the correct methodology. as its use to determine the effect on nontarget species.

Of the nematodes commonly used as a biological model for the assessment of anthelmintic products in vitro is *Caenorhabditis elegans*, and in vivo assessments have used rodents when the parasite allows it [43–45]. It is essential to consider that both physical and biological differences between nematode species can be a limitation for the assessment of new drugs developed to inhibit specific parasitic stages; therefore, this characteristic should be considered when using a nontarget species. (*C. elegans*, entomopathogenic nematodes, among others) for assessments. Currently, new tests or more specific tests are being published to mitigate the deficiency of not using the target species.

8. Nematocide assessment techniques on phytopathogenic nematodes

Nematocides are usually toxic with a broad spectrum and have high volatility or other properties that promote migration through the soil. The use of chemical nematocides is increasing every day even though they have been banned [46] or even though alternative nematocides are being created [47, 48].

Despite the diversity of methodologies used in the assessment of nematocides, the results are differences between assessed compounds, and the same assessment must be performed several times; sometimes, the limited number of nematodes cultured in laboratory conditions does not allow for the necessary repetitions. In addition, the standardisation time of tests can be prolonged.

There is a diversity in the assessment methodologies of phytopathogenic nematocides. From petri dish assessments with the challenge of the chemical being in relatively large spaces, which allow the nematode much movement [49], to the assessment in cell culture chambers [50] with an incubation and assessment period similar to those carried out in studies on parasitic nematodes of animals [51].

Due to the characteristics of entomopathogenic nematodes, they will always be a biological model with great availability to establish an assessment technique in a new laboratory, since the availability of specimens allows us to test a bioassay a greater number of times before using phytopathogenic nematodes and thus train the personnel who will carry out the process.

9. Advantages of the possible use of entomopathogenic nematodes as a model for the biological assessment of anthelmintics

As in all tests where biological organisms are used, there are advantages and disadvantages, in this case between the test performed and between the organisms used as a biological model, so the main objective is to strengthen and exploit the advantages. Below are some advantages and disadvantages of the possible use of entomopathogenic nematodes as biological models to determine the effectiveness of a biological or chemical nematocide of parasitic nematodes of plants or animals.

Advantages

- Small size.
- The short life cycle is quickly completed.
- In the right conditions they reproduce all year round.
- Short life span.
- Known and simple anatomy.
- Abundant progeny.
- Simple and economical cultivation.
- It can be maintained for long periods in the laboratory.
- Several strains can be easily maintained in a small space.
- Constant motility with little stimulus.
- They are cosmopolitan.
- They live in the rhizosphere in more than one of their stages of the biological cycle.
- Isolation and identification are relatively fast, inexpensive and do not require much training.

Disadvantages

- Relatively simple anatomy.
- Possible problem due to the type of feeding in its different stages of development.
- It is not possible to assess bioavailability and organic toxicity.
- It could only be used for standardisation of tests.
- The toxicity to entomopathogenic nematodes is probably not similar to that of parasitic nematodes.
- The differences in their anatomy and physiology should be considered in studies.

10. Conclusions

Currently, there is an urgent demand for the development of new anthelmintic drugs due to various circumstances; reaching their generation and assessment is not a short route and requires many economic resources; every time there has been the need for them, it has been solved. However, the assessment of the product or the new drug in controlled or field conditions is extremely complicated.

Research is required for continuous improvement in the management of parasitic nematodes, emphasising the reduction of the use of chemotherapeutics and the development of resistance to anthelmintics, for which viable options are required for the assessment of this resistance. Entomopathogenic nematodes offer an opportunity that favours these aspects, in addition to helping to understand the interactions of these chemicals with the rhizosphere and the environment in general once they are excreted by animals.

Complications in the assessment of new drugs can be analysed in various ways; however, this chapter proposes an alternative solution for the lack of target nematodes (human or animal parasites) in sufficient quantity and in the biological stage of the nematodes in which the assessment is desired. Entomopathogenic nematodes, due to their characteristics, are an alternative to perform the assessments of new drugs on nontarget nematodes that even allow generating populations resistant to a chemical product to assess a new drug, combinations of them or to simulate activities that limit dispersion of the resistance.

Conflict of interest

The authors declare no conflict of interest.

Author details

Oscar Barrón-Bravo[1], Ismael Montiel-Maya[2], Ana Cruz-Avalos[3], Fidel Avila-Ramos[2], Jaime Molina Ochoa[4] and César Angel-Sahagún[2]*

1 Las Huastecas Experimental Field-National Institute of Forestry, Agricultural and Livestock Research, Villa Cuauhtémoc, Tamaulipas, Mexico

2 Department of Veterinary and Zootechnics, University of Guanajuato, Irapuato, Guanajuato, Mexico

3 Technological University of Southeast Guanajuato, Valle de Santiago, Guanajuato, Mexico

4 Faculty of Biological and Agricultural Sciences, University of Colima, Mexico

*Address all correspondence to: csahagun@ugto.mx

References

[1] Kiontke, K., Fitch, D., 2013. Nematodes. Current Biology. 23: r862-r864.

[2] Seesao, Y., Gay, M., Merlin, S., Viscogliosi, E., Aliouat, D., Audebert, C., 2017. A review of methods for nematode identification. Journal of Microbiological Methods. 138: 37-49.

[3] Rose, G.A., 2018. The fish nematode problem in major European fish stocks. Fisheries Research. 202: 1-3.

[4] Lacey, L.A., 2012. Manual of Techniques in Invertebrate Pathology. Second Edition. Second edition. Elsevier Ltd. Yakima, Washington, USA. p. 384.

[5] Bird, A.F., Bird, J., 1991. The Structure of Nematodes. Academic Press, Second Edition. p. 96-120.

[6] Cordero del Campillo, M., Rojo, F., Martínez, A., Sánchez, C., Hernández, S., Navarrete, J., Díez, P., Quiroz, H., Carvalho, M., 1999. Parasitología Veterinaria. Ed. Mc Graw Hill Interamericana. Madrid, España.

[7] Quiroz-Romero, H., 2000. Parasitología y enfermedades parasitarias de animales domésticos. Editorial Limusa. p. 16-43.

[8] Risser, K., Greenwood, C., Walker, N., Payton, M., Talley, J., 2016. Prevalence and Diversity of Entomopathogenic Nematodes Spanning a Mean Annual Precipitation Gradient in Pastureland in Oklahoma. Southwestern Entomologist. 41 (4): 933-944.

[9] Bedding, R.A., Akhurstr, R.J., 1975. A simple technique for the detection of insect parasitic rhabditid nematodes in soil. Nematologica. 21: 109-116.

[10] Campos-Herrera, R., Blanco-Pérez, R., Bueno-Pallero, F., Duarte, A., Nolasco, G., Sommer, R., Rodríguez, M.J., 2019. Vegetation drives assemblages of entomopathogenic nematodes and other soil organisms: Evidence from the Algarve, Portugal. Soil Biology and Biochemistry 128: 150-163.

[11] McMullen, J.G., Stock, S.P., 2014. In vivo and In vitro Rearing of Entomopathogenic Nematodes (Steinernematidae and Heterorhabditidae). Journal of Visualized Experiments. 91: 1-7.

[12] Canul-Ku, H., Rodríguez-Vivas, R., Torres-Acosta, J., Aguilar-Caballero A., Pérez-Cogollo, A., Ojeda-Chi, M., 2012. Prevalence of cattle herds with ivermectin resistant nematodes in the hot sub-humid tropics of Mexico. Veterinary Parasitology. 183: 292-298.

[13] Bahmani, M., Rafieian, K., Hassanzadazar, H., Saki, K., Ahmad, K., Delfan, B., 2014. A review on most important herbal and synthetic antihelmintic drugs. Asian Pac J Trop Med. 7: S29-S33.

[14] Gokbulut, C., Nolan, A.M., McKellar Q.A., 2002. Plasma disposition, faecal excretion and in vitro metabolism of oxibendazole following oral administration in horses. Research in Veterinary Science. 72: 11-15.

[15] Ceballos, L., Krolewiecki, A., Juarez, M., Moreno, L., Schaer, F., Alvarez, L.I., Cimino, R., Walson, J., Lanusse, C.E., 2018. Assessment of serum pharmacokinetics and urinary excretion of albendazole and its metabolites in human volunteers. PLoS Negl Trop Dis. 12 (1): e0005945.

[16] Gottschall, D.W., Theodorides, V.J. y Wang, R., 1990. El metabolismo de los antihelmínticos de bencimidazol. Parasitology Today, 6 (4), 115-124.

[17] Lanusse, C., Prichard, R., 1993. Clinical Pharmacokinetics and Metabolism of Benzimidazole Anthelmintics in Ruminants. Drug Metab Rev. 25:235-279.

[18] Olliaro, P., Seiler, J., Kuesel, A., Horton, J., Clark, J.N., Don, R., Keizer J., 2011. Potential Drug Development Candidates for Human Soil-Transmitted Helminthiases. PLOS Neglected Tropical Diseases. 5 (6): e1138.

[19] Plante, G.E., Erian, R., Petitclerc, C., 1981. Renal excretion of levamisole. J Pharmacol Exp Ther. 216 (3): 617-623.

[20] Nielsen, P., Rasmussen, F., 1983. Pharmacokinetics of levamisole in goats and pigs. Veterinary Pharmacology and Toxicology. 241-244.

[21] Steel, J.W., 1993. Pharmacokinetics and metabolism of avermectins in livestock. Veterinary Parasitology, 48(1-4), 45-57.

[22] González, C., Sahagún, P., Diez, L., Fernández, M., Sierra, V., García, V., 2009. The pharmacokinetics and metabolism of ivermectin in domestic animal species. The Veterinary Journal. 179: 25-37.

[23] Horvat, A.J.M., Petrovic, M., Babic, S., Pavlovic, D.M., Asperger, D., Pelko, S., Mance, A.D., Kastelan-Macan, M., 2012. Analysis, occurrence and fate of an-thelmintics and their transformation products in the environment. TrAC Trends in Analytical Chemistry 31:61-84.

[24] Karanth, S., 2014. Trichlorfon. Charles River Laboratories, Reno, NV, USA. Elsevier Inc.

[25] Patzschke, K., Pütter, J., Wegner, L.A., Horster, F.A., Diekmann, H.W., 1979. Serum concentrations and renal excretion in humans after oral administration of Praziquantel — results of three determination methods. European Journal of Drug Metabolism and Pharmacokinetics. 4 (3): 149-156.

[26] Sanchez-Bruni, S.F., Jones, D.G., Mckellar, Q.A., 2006. Pharmacological approaches towards rationalizing the use of endoparasitic drugs in small animals. J. vet. Pharmacol. Therap. 29 (6): 443-457.

[27] Hennessy, D.R., Ali, D.N., 1997. The effect of feed intake level on the pharmacokinetic disposition of closantel in sheep. International Journal for Parasitology. 27 (9): 1081-1086.

[28] Lanusse, C., Lifschitz, A., Alvarez, L., 2015. Basic and clinical pharmacology contribution to extend anthelminticmolecules lifespan. Veterinary Parasitology. 212 (1-2), 35-46.

[29] Gokbulut, C., Aksit, D., Smaldonec, G., Mariani, U., Veneziano, V., 2014. Plasma pharmacokinetics, faecal excretion and efficacy of pyrantel pamoate paste and granule formulations following per os administration in donkeys naturally infected with intestinal strongylidae. Veterinary Parasitology. 205: 186-192.

[30] Gokbulut, C., Nolan, A.M., McKellar, Q.A., 2001. Pharmacokinetic disposition and faecal excretion of pyrantel embonate following oral administration in horses. J. Vet. Pharmacol. Ther. 24: 77-79.

[31] Sepúlveda-Crespo, D., Reguera, R.M., Rojo-Vázquez, F., Balaña-Fouce, R., Martínez-Valladares, M., 2020. Drug discovery technologies: *Caenorhabditis elegans* as a model for anthelmintic therapeutics. Med Res Rev. 1-39.

[32] Sumano, H., Ocampo, L., 2006. Farmacología Veterinaria. McGraw-Hill Interamericana Editores. Tercera Edición. p. 472.

[33] Plumb, D.C., 2008. Plumb's Veterinary Drug Handbook. Sixth Edition. p. 303-305.

[34] OMS, 2000. Organización Mundial de la Salud. Evaluación de ciertos residuos de fármacos de uso veterinario en los alimentos. 36° Informe del Comité Mixto FAO/OMS de expertos en Aditivos Alimentarios. Serie de Informes Técnicos 799. ISBN 9243207997. ISSN0509-2507. Impreso en España.

[35] Peachey, L., Pinchbeck, G., Matthews, J., Burden, F., Lespine, A., 2017. P-glycoproteins play a role in ivermectin resistance in cyathostomins. IJP: Drugs and Drug Resistance. 7: 388-398.

[36] Bai, S. H., Ogbourne, S., 2016. Eco-toxicological effects of the avermectin family with a focus on abamectin and ivermectin. Chemosphere. 154: 204-214.

[37] Hernández-Villegas, M.M., Borges-Argáez, R., Rodríguez-Vivas, R.I., Torres-Acosta, J.F., Méndez-González, M., Cáceres-Farfán, M., 2011. Ovicidal and larvicidal activity of the crude extracts from *Phytolacca icosandra* against *Haemonchus contortus*. Veterinary Parasitology. 179: 100-106.

[38] Demeler, J., Küttler, U., El-Abdellati, A., Stafford, K., Rydzik, A., Varaday, M., Kenyon, F., Jackson, F., Vercruysse, J., Von Samson, G., 2010. Standardization of the larval migration inhibition test for the detection of resistance of ivermectin in gastrointestinal nematodes of ruminants. Veterinary Parasitology. 174: 58-64.

[39] McArthur, C., Handel, I., Robinson, A., Hodgkinson, J., Bronsvoort, B., Burden, F., Kaplan, R., Matthews, J., 2015. Development of the larval migration inhibition test for comparative analysis of ivermectin sensitivity in cyathostomin populations. Veterinary Parasitology. 212: 292-298.

[40] Bernt, U., Junkersdorf, B., Londershauesen, M., Harder, A., Schierenberg, E., 1998. Effect of anthelminthics with different modes of action on the behavior and development of Caenorhabditis elegans. Fundamental and Applied Nematology. 21: 251-263.

[41] Mayoral-Peña, Z., Piña-Vazquez, D., Gómez-Sánchez, M., Salazar-Olivo, L., Aguilar-Tipacamú, G., Arellano-Carbajal, F., 2017. The nematode *Caenorhabditis elegans* as a model to assess the anthelmintic potential from plant extracts. Revista Mexicana de Ciencias Pecuarias. 8: 279-289.

[42] Peña-Espinoza, M., López-Muñoz, R., 2018. Caenorhabditis elegans: nematodo modelo para el estudio de compuestos antihelmínticos naturales. Revista Parasitología Latinoamericana. 45.

[43] Risi, G., Aguilera, E., Ladós, E., Suárez, G., Carrera, I., Álvarez, G., Salinas, G., 2019. Caenorhabditis elegans infrared-based motility assay identified new hits for nematicide drug development. Vet Sci. 6(1): 29.

[44] Weeks, J.C., Roberts, W.M., Leasure, C., Suzuki, B.M., Robinson, K.J., Currey, H., Wangchuk, P., Eichenberger, R.M., Saxton, A.D., Bird, T.D., Kraemer, B.C., Loukas, A., Hawdon, J.M., Caffrey, C.R., Liachko, N.F., 2018. Sertraline, paroxetine, and chlorpromazine are rapidly acting anthelmintic drugs capable of clinical repurposing. Scientific Reports 8(1):975.

[45] Partridge, F.A., Brown, A.E., Buckingham, S.D., Nicky, J.W., Graham, M.W., Ruth, F., Kathryn, J.E., Alison, A.M., Jacqueline, B.M., Angela, J.R., David, A.L., David, B.S., 2018. An automated high-throughput system for phenotypic screening of chemical libraries on *C. elegans* and parasitic nematodes. Int J Parasitol Drugs Drug Resist. 8(1): 8-21.

[46] Ghareeb, R.Y., Alfy, H., Fahmy, A.A., Ali, H.M., Abdelsalam, N.R., 2020. Utilization of Cladophora glomerata extract nanoparticles as eco-nematicide and enhancing the defense responses of tomato plants infected by *Meloidogyne javanica*. Scientific Reports. 10: 1-15.

[47] Molinari, S., 2016. Systemic acquired resistance activation in solanaceous crops as a management strategy against root-knot nematodes. Pest management science, 72: 888-896.

[48] Stirling, G.R., 2011. Biological control of plant-parasitic nematodes: an ecological perspective, a review of progress and opportunities for further research. Biological Control of Plant-Parasitic Nematodes, 1-38.

[49] Thoden, T.C., Wiles J.A., 2019. Biological attributes of Salibro™, a novel sulfonamide nematicide. Part 1: Impact on the fitness of *Meloidogyne incognita*, *M. hapla* and *Acrobeloides buetschlii*. Nematology 21 625-639.

[50] Wram, C.L., Zasada, I.A., 2019. Short-Term Effects of sublethal doses of nematicides on Meloidogyne incognita. Phytopathology 109: 1605-1613.

[51] Barrón-Bravo, O.G., Hernández-Marín, J. A., Gutiérrez-Chávez, A.J., Franco-Robles, E., Molina-Ochoa, J., Cruz-Vázquez, C.R., Ángel-Sahagún, C.A., 2020. Susceptibility of entomopathogenic nematodes to ivermectin and thiabendazole. Chemosphere. 253: 126658.

Chapter 12

Entomopathogenic Nematodes: Their Characterization, Bio-Control Properties and New Perspectives

Himani Sharma, Aasha Rana, Aashaq H. Bhat and Ashok K. Chaubey

Abstract

The insect parasitoid nematodes are a means boon to agronomy and serve as important bio-pesticides for controlling crop damaging insect pests. These nematodes inhabit moist soils and have been to exist in all the continents excluding Polar Regions. These nematodes have 3rd larval stage infective which is the only free living stage existing outside the host. These infective stages are mutually associated with bacteria which reside in their alimentary canal and duo are responsible for mortality of the insect host. These nematodes are currently given great attention by scientific community because of their insect killing properties and can be used to replace hazardous pesticides. These nematodes include various species belonging to genus *Heterorhabditis* and *Steinernema*, and members of insectivorous group of genus *Oscheius*. Before their use as bio-control agents, these nematodes need to be properly identified. Currently, these nematodes are characterized by using morphological and morphometrical parameters and advanced molecular tools including cross hybridization and scanning electron microscope studies. Their associated bacterial partners are studied through advanced molecular and biochemical techniques. The properly characterized nematodes having more entomopathogenic properties can be easily mass produced through *in vitro* and *in vivo* methods. They can be formulated in various carrier materials and supplied to farmers for effective control of damaging insect pests. Several countries have formulated various useful products of entomopathogenic nematodes which are available in markets for use by the farmer community and some have given very effective results. India is still at the early stage in the use of these nematodes for bio-control of insects in agronomy. More research in this field needs to be carried, especially in India to produce effective indigenous nematode products which may prove a boon for agriculture.

Keywords: *Steinernema*, *Heterorhabditis*, biological control, and pathogenicity

1. Introduction

1.1 Entomopathogenic nematodes

Entomopathogenic nematodes (EPNs) range in size from 0.3 to 10 mm and they can be more or less cylindrical [1]. In Greek vocables, the term entomopathogenic nematodes comes from "entomos", "insects", "pathê", "disease" and "guenos", "producing" means a group of nematodes which have the ability to cause disease in insects by suppressing the immune system of insects. "Entomopathogenicity clarified: "*EPNs must rapidly kill their hosts with the aid of bacterial partners and must pass on the associated bacteria to future generations*" [2]. They belong to the two families, the Steinernematidae consisting of two genera, *i.e. Steinernema* (100 valid species) and *Neosteinernema* (01 species only, *N. longicurvicauda*) [3]. The other genera Heterorhabditidae comprises of one genus, *Heterorhabditis* which contains 16 well described species globally [3] (**Table 1**). These two well-known genera, *Steinernema* and *Heterorhabditis*, have the ability of infecting and killing insects with the aid of symbiotic bacteria [91]. They are receiving a lot of interest in nematological and entomological studies because of their high virulence capacities, and able to kill the insect hosts within 24–48 hours. Besides this, they are ubiquitous and reside everywhere except Antarctica [76, 115–119].

S. No	Species	Place	Reference	S. No	Species	Place	Reference
1	*S. kraussei*	Germany	[4]	60	*S. ichnusae*	Italy	[5]
2	*S. glaseri*	New Jersey	[6]	61	*S. australe*	Chile	[7]
3	*S. feltiae*	Russia	[8]	62	*S. unicornum*	Chile	[9]
4	*S. affine*	Denmark	[10, 11]	63	*S. boemarei*	France	[12]
v5	*S. carpocapsae*	Czechoslovakia	[13]	64	*S. xueshanense*	China	[14]
6	*S. intermedium*	Carolina, USA	[15]	65	*S. brazilense*	Brazil	[16]
7	*S. rarum*	Córdoba, Argentina	[17]	66	*S. schliemanni*	Germany	[18]
8	*S. kushidai*	Shizuoka, Japan	[19]	67	*S. minutum*	Thailand	[20]
9	*S. ritteri*	Córdoba, Argentina	[21]	68	*S. arasbaranense*	Iran	[22]
10	*S. scapterisci*	Uruguay	[23]	69	*S. citrae*	South Africa	[24]
11	*S. caudatum*	China	[25]	70	*S. nepalense*	Nepal	[26]
12	*S. neocurtillae*	Florida, USA	[27]	71	*S. surkhetense*	Nepal	[28]
13	*S. longicaudum*	China	[29]	72	*S. lamjungense*	Nepal	[26]
14	*S. cubanum*	Cuba	[30]	73	*S. phyllophagae*	Florida, USA	[31]
15	*S. riobrave*	Texas, USA	[32]	74	*S. pui*	China	[33]
16	*S. puertoricense*	Loiza, Puerto Rico	[34]	75	*S. changbaiense*	China	[35]
17	*S. bicornutum*	Serbia	[36]	76	*S. ethiopiense*	Ethiopia	[37]
18	*S. oregonense*	Oregon, USA	[38]	77	*S. tielingense*	China	[39]
19	*S. abbasi*	Sultanate of Oman	[40]	78	*S. xinbinense*	China	[41]
20	*S. arenarium*	Russia	[42]	79	*S. cameroonense*	Cameroon	[43]
21	*S. ceratophorum*	China	[44]	80	*S. nyetense*	Cameroon	[43]
22	*S. monticolum*	Korea	[45]	81	*S. sacchari*	South Africa	[46]
23	*S. karii*	Kenya	[47]	82	*S. tophus*	South Africa	[48]
24	*S. siamkayai*	Thailand	[49]	83	*S. huense*	Vietnam	[50]
25	*S. tami*	Vietnam	[51]	84	*S. poinari*	Czech Republic	[52]

S. No	Species	Place	Reference	S. No	Species	Place	Reference
26	*S. loci*	Vietnam	[53]	85	*S. innovation*	South Africa	[54]
27	*S. sangi*	Vietnam	[55]	86	*S. jeffreyense*	South Africa	[56]
28	*S. thanhi*	Vietnam	[53]	87	*S. papillatum*	Venezuela	[57]
29	*S. pakistanense*	Karanchi, Pakistan	[58]	88	*S. beitlechemi*	South Africa	[59]
30	*S. asiaticum*	Pakistan	[60]	89	*S. pwaniensis*	Tanzania	[61]
31	*S. diaprepesi*	Florida	[62]	90	*S. fabii*	South Africa	[63]
32	*S. anatoliense*	Turkey	[64]	91	*S. nguyeni*	South Africa	[65]
33	*S. scarabaei*	New Jersey, USA	[66]	92	*S. biddulphi*	South Africa	[67]
34	*S. weiseri*	Czech Republic	[68]	93	*S. ralatorei*	Mexico	[69]
35	*S. apuliae*	Italy	[70]	94	*S. litchi*	South Africa	[71]
36	*S. guangdongense*	China	[72]	95	*S. borjomiense*	Georgia, USA	[73]
37	*S. hermaphroditum*	Indonesia	[74]	96	*S. khuongi*	Florida, USA	[75]
38	*S. jollieti*	USA	[76]	97	*S. taiwanensis*	Taiwan	[77]
39	*S. litorale*	Japan	[78]	98	*S. bertusi*	South Africa	[79]
40	*S. yirgalemense*	Ethiopia	[80]	99	*S. riojaense*	Spain	[81]
41	*S. aciari*	China	[82]	100	*S. sandneri*	Poland	[83]
42	*S. akhursti*	China	[84]	101	*S. batswanae*	South Africa	[85]
43	*S. beddingi*	China	[86]	102	*H. bacteriophora*	Australia	[87]
44	*S. robustispiculum*	Vietnam	[88]	103	*H. megidis*	USA	[89]
45	*S. silvaticum*	Germany	[90]	104	*H. zealandica*	South Africa	[91]
46	*S. ashiuense*	Japan	[92]	105	*H. indica*	India	[93]
47	*S. backanense*	Vietnam	[94]	106	*H. marelatus*	Oregon, USA	[95]
48	*S. cumgarense*	Vietnam	[94]	107	*H. taysearae*	Egypt	[96]
49	*S. sasonense*	Vietnam	[94]	108	*H. downesi*	Ireland	[97]
50	*S. eapokense*	Vietnam	[94]	109	*H. baujardi*	Vietnam	[98]
51	*S. khoisanae*	South Africa	[99]	110	*H. maxicana*	Mexico	[100]
52	*S. leizhouense*	China	[101]	111	*H. amazonensis*	Brazil	[102]
53	*S. sichuanense*	China	[103]	112	*H. floridensis*	Florida, USA	[104]
54	*S. hebeiense*	China	[105]	113	*H. Georgiana*	Georgia, USA	[106]
55	*S. costaricense*	Costa Rica	[107]	114	*H. safricana*	South Africa	[108]
56	*S. puntauvense*	Costa Rica	[107]	115	*H. atacamensis*	Chili	[109]
57	*S. texanum*	Texas	[110]	116	*H. beicherriana*	China	[111]
58	*S. cholashanense*	China	[112]	117	*H. noenieputensis*	South Africa	[113]
59	*S. colombiense*	Colombia	[114]				

Table 1.
List of valid Steinernema and Heterorhabditis species identified worldwide with geographical location and respective authority.

1.2 Life cycle of EPNs

Steinernema and *Heterorhabditis* share the similar life cycles. Both genera balance between a free-living stage and a parasitic stage (**Figure 1A-J**). The free-living form of EPNs is protected from the environment by an external cuticle. Being encapsulated, the invasive EPN stage, called infective juvenile (IJ) corresponding to J3, are unable to feed because their mouth and anus are sealed [120]. They rather possess huge lipid storage to be able to survive outside a host for several months [121]. With comparable lipid reserves, it has been shown that the IJs of *Steinernema* survive longer in the environment than *Heterorhabditis* IJs, which can be explained by the motile behavior of IJs. It has been found that IJs of *Steinernema* nictate between 50 and 80% of their life time while *Heterorhabditis* IJs nictate between 70 and 90% [122]. As a result of which lipid reserves are consumed faster in the IJs of *Heterorhabditis* as compared to *Steinernema*. These IJs wait for insect larvae up to 20 cm deep in soil [123]. In case of *Steinernema*, IJs invade the insect larvae through natural openings such as the mouth, anus, spiracles and wounds [124]. However, in case of *Heterorhabditis,* the IJs penetrate the insect body by directly scratching their cuticle as they are equipped with a large anterior tooth [125, 126]. Once inside the host, IJs lose their cuticle and release the entomopathogenic bacteria (EPB) and this nematode- bacterium complex together is lethal for the insect host.

The IJs feed on the dead insect cadaver and mature into the fourth stage juveniles (J4) which differentiate into males and females, generally 3 days post insect infestation. After mating, the first generation (G1) females lay eggs, either in the external

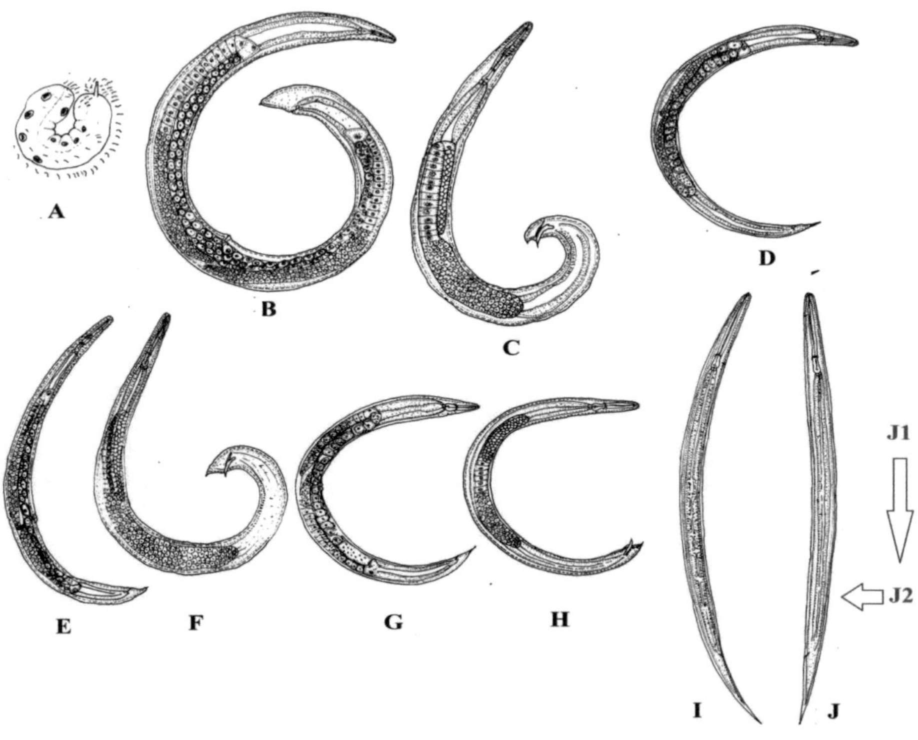

Figure 1.
A-Nematodes enters into host insect; B,C- First generation female of Steinernematidae; D- First generation hermaphrodite female of Heterorhabditidae; E,F-Second generation female and male of Steinernematidae; G,H-Second generation of Amphimictic female and male of Heterorhabditidae; I,J-Infective juvenile (IJ) stage of Heterorhabditidae and Steinernematidae.

medium or remaining in the maternal body, which hatch into the first-stage juveniles (J1). At that point, two scenarios are possible depending on the amount of food available in the insect cadaver. In case of scarce food, J1 molts into the second-stage juvenile (J2) within 2 or 3 days. Then J2 ceases to feed and molts into pre-infective stage juvenile, also called immature IJs, before becoming infective juvenile. Then the newly generated IJ emerge from the depleted insect cadaver to actively look for another susceptible insect prey. On the contrary, if food is abundant in the cadaver, then several generations of males and females can be produced in the same cadaver. After hatching from the G1 females' eggs, J1 molt successively into J2, non-infective J3 and J4 developing into the second generation (G2) adults. After mating, G2 females produce eggs that mature into J1, thereby initiating a new cycle. EPNs usually reproduce 2 or 3 generations before total depletion of the food resources in the insect cadaver occurs [124]. The entire reproductive cycle lasts between 7 and 14 days, mainly depending on temperature, after insect invasion by IJs. Both *Steinernema* and *Heterorhabditis* females lay eggs in the insect cadaver after mating with males. Juveniles hatched from released eggs often develop into amphimictic adults [127].

The reproductive life cycle of most *Steinernema* involves both sexually differentiated partners, G1 males and females whilst all *Heterorhabditis* IJs develop into self-fertilizing hermaphrodite females after insect infection [91]. However the second generation produces amphimictic *Heterorhabditis* adults. Interestingly, IJs from the species of *S. hermaphroditum* can develop into self-fertilizing hermaphrodite females just like *Heterorhabditis* IJs do. It has been argued that the uncommon feature of this *Steinernema* species supports the independent but convergent evolution with *Heterorhabditis* postulated by Poinar and described before [128]. As a consequence of the hermaphrodite reproduction of *Heterorhabditis* EPNs, the genetic diversity of offspring is highly decreased or impaired. The hermaphrodite behavior of *Heterorhabditis* allows infection by a single IJ molting into a hermaphrodite female while at least two *Steinernema* IJs have to invade an insect larva and develop into male and female [129]. Certainly, this provides a real advantage to the survival of *Heterorhabditis* species over *Steinernema* species.

Mating between males and females consists in introducing sperm to fertilize the female's eggs. Male introduces its spicule to the vulva of female and produces spermatozoids and release them in vulva. The male's sperm fertilizes female's eggs in the uterus. For hermaphrodites, sperm is produced and stored into the spermatic vesicles described as distal swelling of the uterus. When the female starts laying eggs, they are automatically fertilized by the sperm contained within the spermatic vesicles [127, 130]. Since the females are larger in size, males have to find a way to scan the entire female body to be able to find the vulva. Male finds the vulva of the female body by the two ways. These two reproductive behaviors point out another distinction between *Steinernema* and *Heterorhabditis i. e* males stick to a female and slips all along the female body until it finds the vulva *viz.* both *Heterorhabditis* female and male heads are pointing in opposite direction [131]. The males act like a ring around the female body *viz. Steinernema*. The male coils around and all along the female body until it reaches the vulva [132]. Some mechanisms do exist to avoid several males copulating with the same female. In *Heterorhabditis* species, male leaves a mating plug closing the vulva after mating preventing other males to mate with the same female [93]. In *Steinernema* species, it has been shown that virgin females produce some chemical attractants for males and their production decreases after mating [132]. However, in *S. longicaudum*, males need the presence of virgin conspecific females in their close environment [133].

After mating, a lot of eggs are retained inside the EPN maternal body, offspring hatch and start feeding inside their maternal body. This phenomenon is known as *endotokia matricida*, the term comes from two Greek words "endo", inside and "tocos", birth and two Latin words "mater", mother and "caedere", kill. *Endotokia matricida* promotes in the scarce food condition supply, then, this condition occurs for the first generation of juveniles. It becomes then obvious that the size of the susceptible insect will affect the development and survival of EPNs. Few authors reported the inefficiency of *Steinernema* IJs to control micro-insect pests [134, 135]. *Steinernema*, and *Heterorhabditis*, nematodes cannot persist for a long time in the environment if no larger insects are available to them for completing their life cycle [136].

1.3 Nematode movement and host location

The 3rd stage infective juveniles of *Steinernema* and *Heterorhabditis* move freely in soil in search of the host and have been distinguished into three categories on the basis of their host finding behaviors-: (i) cruisers- species whose IJs actively move through a substrate to find a host (ii) ambushers- species that employ a '*sit and wait*' strategy that involves little displacement and active searching and (iii) intermediates- show both the types of behaviors [137]. All *Heterorhabditis* species are cruisers [138]; however, *Steinernema* genus displays all three behaviors'. *S. carpocapsae* displays ambush behavior and *S. feltiae* shows intermediate behaviors [132]. Some ambushers have the characteristic feature to stand on the substrate. At the time of standing, IJs raise a portion of the anterior section of their body off the substrate, sometimes waving it back and forth, a process referred to as 'nictation'. This process of nictation and standing is of varying duration, commonly observed in *S. carpocapsae* and showing this phenomenon over protracted periods of time [139]. Besides this, in some species of *Steinernema,* IJs have been observed to jump which helps them in traveling longer distances [140]. This jump behavior is utilized in dissemination and might sometimes serve as a search mechanism of these EPNs to attack at passing hosts [137, 141, 142]. This jumping and/or standing behavior is advantageous in ambushers to disperse easily and bridge large pores found in some substrates (loose, porous soils or organic litter) than cruisers that do not nictate and instead move across the surface of soil particles [140, 143]. As far as cruisers are concerned, they are thought to be attracted towards host by the host volatiles and host cues from a distance [144]. Ambushers like *S. carpocapsae* do not show any change in behavior to host cues, while cruisers like *S. glaseri* does show varying behaviors [145, 146].

Only 3rd stage of EPNs is considered as infective and pathogenic which is called the infective juvenile (IJ). Infective juveniles are the only free-living stage of EPNs, while other developmental stages are only found inside infected insect hosts. The IJs are stress tolerant, non-feeding, bacterial vectoring stages that seeks out insects to infect and kill. The IJs penetrate the host insect either through natural openings like spiracles, mouth, and anus or in some species through intersegmental membranes of the cuticle, and then enter into the homocoel [110, 125, 147]. *Heterorhabditis* species apart from natural openings also penetrate the insect host by abrading the skin. After penetrating into the skin, the IJs release cells of their symbiotic bacteria from their alimentary canal into the hemocoel (**Figure 2**). The bacteria multiply in the insect hemolymph, secrete toxins and targeted immune depressors that suppress the insect's immune system resulting in death with 24–48 hours [148, 149]. *Photorhabdus* and *Xenorhabdus*, two well-known bacterial symbionts of EPNs, are not only lethal to entomic fauna but also prevent opportunistic bacteria and fungi from utilizing the nutrient rich cadaver, sequestering

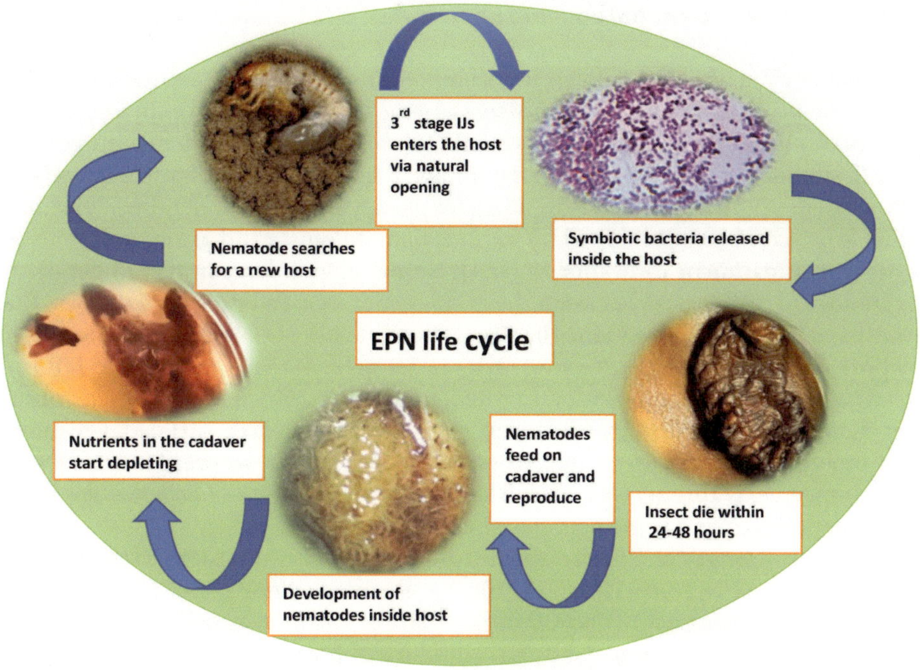

Figure 2.
Life cycle of the Entomopathogenic nematodes inside the host insect.

the resources for themselves and their nematode partners. The pathogenic bacteria always contribute to the virulence of the duo, and usually contribute the lion's share. In some species, nematodes are believed to serve as carriers of bacteria and play very little role in the death of the host, while in others, nematodes are itself responsible for the death of the host by secreting a variety of protein products that degrade and digest the host tissues, in addition to weakening the host immune system. Till date, no nonbacterial associated EPNs are known to science even if some nematodes appear lethal on their own. The nematodes, after the death of the host, continue to feed upon the bacteria and liquefy cadaver tissues, develop into mature adults (male and female) and reproduce. If there is plenty of food, the IJs develop into second generation adults and continue their life cycle. One or more generations may develop within the cadaver depending on the availability of food resources and once the food is depleted in the cadaver, a large number of IJs are eventually released into the environment to infect other hosts and continue their life cycle [150–152]. The IJs can live for weeks on stored food reserves and for months by entering a near-anhydrobiotic state. Their persistence in the soil depends on two key features *i.e.* the availability of an insect host and their progeny production in that host.

The process of reproduction in heterorhabditid and steinernematid nematodes shows few differences. The IJ of steinernematids develop into amphimictic males and females in all the adult generations (gonochorisism) while in heterorhabditid nematode IJ develop into self-fertilizing hermaphrodites in the first generation and in second generation, produce males, females and hermaphrodites [153]. The insect cadaver becomes red if the insects are killed by heterorhabditids and brown or tan if killed by Steinernematids [150]. The color of the host body is indicative of the pigments produced by the monoculture of mutualistic bacteria growing in the hosts.

1.4 Taxonomy of entomopathogenic nematodes

Morphology is one of the major key components of classical taxonomy. It briefs out the genetic organization of organisms as genes themselves are expressed in the form of phenotype. *Steinernema* and *Heterorhabditis* are closely related genera under Rhabditida. Based on their morphology, they are very similar to each other, making them undistinguishable for a non-expert eye. However, detailed systematic feature keys have been developed by various nematologists and are currently used for the identification of EPN species [154]. Nematologists have provided detailed morphological differences between these two genera, including their families' that are *i.e.* used in their precise identification. These include: (i) position of excretory pore (EP) anterior to nerve ring (NR) in *Steinernema* and posterior to NR in *Heterorhabditis,* (ii) color variation in infected cadavers which appears black or no color change in *Steinernema,* while brick red in *Heterorhabditis*, (iii) cadaver showing of bioluminescence when infected with *Heterorhabditis*, whereas in *Steinernema* bioluminescence is absent, (iv) *Heterorhabditis* associated with *Photorhabdus* bacterial partners and *Steinernema* associated with *Xenorhabdus* [155].

Based on the length of IJs four 'species groups' have been created: *glaseri group* (IJ >1000 μm); *feltiae group* (IJ = 700–1000 μm); *intermedium group* (IJ = 600–700 μm); *corpocapsae group* (IJ <600 μm). Another group '*bicornutum*' have also been created, which is diagnosed by the presence of horn like structures on their labial region. The male reproductive apparatus spicule is the most discriminative features in identification of steinernematids, however in second generation males, spicules are more separated from each other [156, 157].

Adults (1st and 2nd generations) and IJs of *Steinernema* and *Heterorhabditis* show some distinctive morphological features which are important for the taxonomic point of view. These characteristics are tail length; position of excretory pore (EP) and nerve ring (NR), pharynx and neck length (PL), beside these, male acquires spicule and gubernaculum. The SEM studies of the 1st generation males reveal the comprehensive structure of gubernaculums and spicules [123, 158]; presence or absence of caudal mucron, disposition of the copulatory papillae, spermatozoon morphology [159] and presence or absence of small cuticular projections *i.e.* the epiptygmata, guarding the entrance of the of the female vagina. In case of IJs, lateral field, tail shape and length, head contour, cephalic horns etc. are some of the important characteristic of taxonomic importance [160]. Measurements and analysis of these characters play an important role in proper identification of EPNs. For example, structure of vulva gives the taxonomists a comprehensible way in recognition of species.

Now-a-days, morphological characterization does not give reliable outcomes as there has been an increase in the number of species which makes the molecular characterization mandatory for the identification of species. Morphology is entirely dependent on the external features of the specimen; however, some genes have the tendency to not express themselves in the form of phenotype although they possess some conserved regions which are very important from the taxonomic point of view. Furthermore, morphology is a tedious task and requires good skilled taxonomists with the expertise in this area. This creates a demand for the molecular identification and validation of a particular species. Advancements in the molecular techniques help in the precise identification and placement of the species in its appropriate position in the classification. A number of molecular techniques are being used for more precise identification of EPNs like immunological techniques [161]; isoenzyme patterns [162]; total protein patterns [6] and RFLP detection within total genomic DNA [163–165]. Nowadays, regions of taxonomic importance which include the internal transcribed spacer (ITS) of the ribosomal DNA (rDNA) repeat unit, 18S and 28S rDNA and the cytochrome oxidase subunit II (COII) are widely used for nematode identification [166–169].

With the advancement in molecular identification, techniques like polymerase chain reaction, amplification and sequencing of the amplified products of the conserved areas became possible. 28S- and 18S rDNA are used compare the distant taxa that had diverged a long time ago. Besides this, IGS, ITS1, ITS2 and ETS are being used to compare the phylogeny of closely related species as compared to 28S and 18S rDNA genes [170]. D2D3 is highly variable expansion segment of 28S rDNA, which have been used for molecular taxonomy and phylogenetic relationship of the nematodes species [171]. 18S ribosomal DNA sequences are used to find out the unknown as well as new species of the nematodes by correlating sequence variations with the genetic differences among the nematode populations [172]. Comparison of the small ribosomal RNA (18S rRNA) nucleotide sequence allows distinguishing steinernematidae from heterorhabditidae [173–175]. Due to its high variability, the ITS sequence lying between the 18S and 28S rRNA genes can be used to distinguish between *Steinernema* and *Heterorhabditis* at the species level [176, 177]. However, ITS sequence analysis is not always sensitive enough and other molecular markers may be required for better identification. The 28S rRNA gene [178], the mitochondrial cytochrome oxidase II (COII)-16S rDNA region and the ND4 mitochondrial gene have been used for that purpose [175, 177, 179].

2. Entomopathogenic nematodes as bio-agents against insect pests

India is a power house of agriculture and has made a great improvement in agriculture, but the crops are damaged by more than 10, 000 species of insects, 30, 000 species of weeds, 1, 00, 000 diseases (caused by fungi, viruses, bacteria and other microorganisms) and 1, 000 species of nematodes [180, 181]. To reduce global crop losses, it has been estimated that around US $ 40 billion are used annually worldwide for the application of 3 million metric tons of pesticides, plus the use of various biological and other non-chemical controls worldwide [182, 183]. Out of total 70,000 estimated pests destroying 35–40% crops, insects are contributing around 14% [183]. To feed a large population of our country, the surge for production of horticultural crops is increasing day by day, due to indiscriminate, unfettered, nonjudicious and rampant use of chemical pesticides and fertilizers and without their use, it is very likely that pests would consume higher percentage and cause huge losses to productivity. A recent United Nations report (2017) assessed that 2, 00, 000 people across the world die per year from toxic exposure of pesticides and cancer problems are increasing from past few years which are directly or indirectly linked to pesticide poisoning (https://www.aljazeera.com/news/2017/03/200000-die-year-pesticide-poisoning-170308140641105.html). Currently, agronomists search for alternate approaches of pest control which are eco-friendly and cost effective like the use of biocontrol agents. One of the earliest examples of classical biological control targeting an insect pest in an agricultural setting is the use of the vedalia beetle, *Rodolia cardinalis*, which was introduced to citrus groves in California from Australia in the late 19th century to counteract the cottony cushion scale, *Icerya purchasi* [184]. Since then, biological control organisms such as fungi, bacteria and EPNs have been used against various insect pests [185–187]. EPNs are important biological control agents and their potential as alternatives to chemical pesticides for controlling pesky insects was recognized early on and they have been subjected to extensive laboratory and field testing [188]. EPNs are safe to most non-target organisms and the environment, are easy to apply, and are compatible with most agricultural chemicals [149]. They also have a broad host range, ability to search for pests, and a potential to reproduce after application [149]. EPN based formulations are commercially available for pest control in home gardens and are commonly marketed as 'beneficial nematodes'. Several species of EPNs were evaluated for their pathogenicity against different pests like *Heterorhabditis bacteriophora* was noticed good control agent for controlling *Ceratiti scapitata* [189] while *H. zealandica* was tested for its ability to control *Planococcus citri*, the citrus mealy bug [190–192].

The species specific EPNs are being used worldwide as biocontrol agents under different trade names *viz*. Ecomask, Savoir Weevil larvae, Guardian, J-3 Max, Heteromask, Lawn Patrol, Scanmask, Entonem, Nemasys etc. and have provided excellent results against the entomic fauna. In United Kingdom and Europe, Bionema company and E-nema company respectively are commercially producing formulations of EPNs so as use them in biological control of different pests and earn millions of US $ every year.

3. The future of EPN systematics: integrating molecules and morphology

Important contribution by various workers seems to be low because nematodes belong to the phylum which is taxonomically, ecologically and geographically diverse group. Nematodes usually comprise 90% of metazoan fauna and a very large number of these creatures are waiting for their discovery. Because the number of species is far from the identified species, progress in this field is still continue and new species are being added but it need tremendous research effort to know the "monopolized kingdom of nematode very well". Lack of adequate taxonomic expertise and non-availability of literature on various described species have been major constraints to identify the species of nematode parasites of insects [193].

The taxonomy of EPNs using molecular tools has made EPN systematics a lot more exciting, and probably will continue to do so in future. The rapid development of molecular techniques promoted the description of several new species and has become the technique of choice for diagnosing EPNs [194]. But morphological investigation too is important and therefore, it would be a mistake to replace traditional (morphological) methods with molecular techniques. The better procedure therefore, is the use of combination of both the approaches which offers a more resourceful perspective for resolving a variety of questions in nematode taxonomy, and particularly for EPNs. The molecular tools should be supplemented with morpho-taxometrical and hybridization tests for validation of a new species. It was found that the combined dataset of molecular and morphology represented the best working of evolutionary history for *Steinernema* [180]. It has been suggested that most morphological features are not phylogenetically informative because they represent plesiomorphic (ancestral) states or are highly homoplasious (caused by convergent or parallel evolution) [195]. For example, presence of less than 8 ridges in the lateral field of infective juveniles, a feature that has been emphasized as indicator of relationship among species, represents an ancestral state. Likewise, the absence of an epiptygma in the 1st generation female vulva is also a plesiomorphic state. Other features such as presence/absence of a tail mucron, spicule morphology, or presence of a velum in the male spicules, were depicted as highly homoplasious. Only two features, presence/absence of lamina notch and presence/absence of tail spines, had significance from a phylogenetic perspective as they were depicted as autapomorphies (unique derived characters). Most nematologists preferred molecular and morphological tools to be complementary tools in EPN systematics. Both approaches present advantages and disadvantages, and also reflect different evolutionary mechanisms, but together provided a more comprehensive view of EPN evolution. The best approach to studying the relationship between EPN species and to determine new species is to integrate both morphological and molecular techniques [76, 196].

Acknowledgements

The authors are thankful to Department of Science and Technology (DST), New Delhi for providing financial assistance through DST WOS-A (SR/WOS-A/LS-1083/2014) to Aasha and DST Inspire Fellowship/2014/76 to Aashaq Hussain Bhat.

Conflict of interest

"The authors declare no conflict of interest."

Author details

Himani Sharma[1], Aasha Rana[1,3]*, Aashaq H. Bhat[1,2] and Ashok K. Chaubey[1]

1 Nematology Laboratory, Department of Zoology, Chaudhary Charan Singh University, Meerut, Uttar Pradesh, India

2 Department of Zoology, Government Degree College, Uttersoo, Anantnag, Jammu and Kashmir, India

3 Department of Environmental Sciences, Chaudhary Charan Singh University Meerut, India

*Address all correspondence to: aasha.aasharana@ymail.com

References

[1] Flint ML, Dreistadt SH. Natural enemies' handbook, the illustrated guide to biological pest control. 1998. pp.2-35.

[2] Dillman AR, Guillermin ML, Lee JH, Kim B, Sternberg PW, Hallem EA. Olfaction shapes host-parasite interactions in parasitic nematodes. PNAS USA. 2012;109(35):E2324-E2333. PubMed: 22851767.

[3] Bhat AH, Chaubey AK, Askary TH. Global distribution of entomopathogenic nematodes, Steinernema and Heterorhabditis. Egypt J Biol Pest Control 30, 31 (2020). https://doi.org/10.1186/s41938-020-0212-y

[4] Steiner G. *Aplectana kraussei* n. sp., eine in der Blattwespe Lyda sp. parasitier ende Nematoden form, nebst Bemerkungen über das Seitenorgan der parasitischen Nematoden. Zentralblatt für Bakteriologie, Parasitenkunde, Infektions krankheiten und Hygiene, Abteilung II . 1923;59:14-18.

[5] López-Núñez JC, Plichta K, Góngora-Botero CE, Stock SP. A new entomopathogenic nematode, *Steinernema colombiense* n. sp. (Nematoda: Steinernematidae) from Colombia. Nematol. 2008;10:561-574.

[6] Poinar GO, Kozodai EM. *Neoaplectana glaseri* and *N. anomali*: sibling species or parallelism? Revue de Nematologie. 1988;11:13-19.

[7] Tarasco E, Mráček Z, Nguyen KB, Triggiani O. *Steinernema ichnusae* sp. n. (Nematoda: Steinernematidae) a new entomopathogenic nematode from Sardinia Island (Italy). J. Invertebr. Pathol. 2008;99(2):173-185.

[8] Glaser RW, Fox H. A nematode parasite of the Japanese beetle (*Popillia japonica* Newm.). Sci. 1930;71:16-17.

[9] Edgington S, Buddie AG, Tymo L, Hunt DJ, Nguyen KB, France AI, Merino LM, Moore D. *Steinernema australe* n. sp. (Panagrolaimomorpha: Steinernematidae), a new entomopathogenic nematode from Isla Magdalena, Chile. Nematol. 2009;11:699-717.

[10] Filipjev IN. Miscellanea Nematologica 1. Eine neue Art der Gattung *Neoaplectana* Steiner nebst Bermerkungen über die systematische Stellung der letzteren. Parazitologichesky. 1934;4:229-240.

[11] Bovein P. Some types of association between nematodes and insects. Videnskabelige Meddelelser Fra Dansk Naturhistorisk Forening, Kobenhavn, 1937;101:1-114.

[12] Edgington S, Buddie AG, Tymo L, France AI, Merino LM, Hunt DJ. *Steinernema unicornum* sp. n. (Panagrolaimomorpha: Steinernematidae), a new entomopathogenic nematode from Tierra del Fuego, Chile. J. Nematode Morphol. System. 2009;12:113-131.

[13] Weiser J. *Neoaplectana carpocapsae* n. sp. (Anguillulata: Steinernematinae), novy cizopasník housenek obaleče jablečného, *Carpocapsa pomonella* L. Věstník Československé Zoologické Společynosti. 1955;19:44-52.

[14] Lee MM, Sicard M, Skeie M, Stock SP. *Steinernema boemarei* n. sp. (Nematoda: Steinernematidae), a new entomopathogenic nematode from southern France. Syst. Parasitol. 2009;72:127-141.

[15] Poinar GOJr. *Neoaplectana intermedia* n. sp. (Steinernematidae: Nematoda) from South Carolina. Nematol. 1986;8:321-327.

[16] Mrácek Z, Liu QZ, Nguyen KB. *Steinernema xueshanense* n. sp.

(Rhabditida, Steinernematidae), a new species of entomopathogenic nematode from the province of Yunnan, southeast Tibetan Mts., China. J. Inver. Path. 2009;102: 69-78.

[17] Doucet MMA. A new species of *Neoaplectana* Steiner, 1929 (Nematoda: Steinernematidae) from Cordoba, Argentina. Nematol. 1986;9:317-323.

[18] Nguyen KB, Ginarte CMA, Leite L, Santos JM, Harakava R. *Steinernema brazilense* n. sp. (Rhabditida: Steinernematidae), a new entomopathogenic nematode from Mato Grosso, Brazil. J. Inver. Path. 2010;103:8-20.

[19] Mamiya Y. *Steinernema kushidain*. sp. (Nematoda: Steinerne-matidae) associated with scarabaeid beetle larvae from Shizuoka, Japan. Appl. Entomol. Zool. 1988;23:313-320.

[20] Spiridonov SE, Waeyenberge L, Moens M. *Steinernema schliemanni* sp. n. (Steinernematidae; Rhabditida) – a new species of steinernematids of the 'monticolum' group from Europe. Russ. J. Nematol. 2010;12:175-190.

[21] Doucet MMA, Doucet ME. Description of *Steinernema ritteri* n. sp. (Nematoda: Steinemematidae) with a key to the species of the genus. Nematologica. 1990;36:257-265.

[22] Maneesakorn P, Grewal PS, Chandrapatya A. *Steinernema minutum* sp. nov. (Rhabditida: Steinernema): a new entomopathogenic from Thailand. Int. J. Nematol. 2010;20:27-42.

[23] Nguyen KB, Smart GC Jr. Addendum to the morphology of *Steinernema scapterisci*, J. Nematol. 1992;24:478-481.

[24] Nikdel M, Niknam GR, Ye W. *Steinernema arasbaranense* n. sp. (Nematoda: Steinernematidae), a new entomopathogenic nematode from Arasbaran forests, Iran. Nematol. Med. 2011;39:17-28.

[25] Xu Z, Wang G, Li X. A new species of the genus *Steinernema* (Rhabditida: Steinernematidae). Zoo. Res. 1991;12:17-20.

[26] Stokwe NF, Malan AP, Nguyen KB, Knoetze R, Tiedt L. *Steinernema citrae* n. sp. (Rhabditida: Steinernematidae), a new entomopathogenic nematode from South Africa. Nematol. 2011;13:569-587.

[27] Nguyen K, Smart G. Steinernema neocurtillis n. sp. (Rhabditida: Steinernematiclae) and a Key to Species of the Genus Steinernema. J Nematol. 1992;24(4):463-77.

[28] Khatri-Chhetri HB, Waeyenberge L, Spiridonov S, Manandhar HK, Moens M. Two new species of *Steinernema* Travassos, 1927 with short infective juveniles from Nepal. Russ. J Nematol. 2011;19:53-74.

[29] Shen CP, Wang GH. Description and studies of an entomopathogenic nematode: *Steinernema longicaudum* sp. nov. In: Proceedings of the first national academy symposium of young and middle aged science and technology workers on plant protection, Beijing, China. Chinese Science and Technology Press, Beijing. 1992; 220-231.

[30] Khatri-Chhetri, Hari B, Waeyenberge L, Moens M, Spiridonov S, Manandhar HK. *Steinernema lamjungense* n. sp. (Rhabditida: Steinernematidae), a new species of entomopathogenic nematode from Lamjung district, Nepal. Nematol. 2011;13(5):589-605.

[31] Mráček Z, Hernandez EA, Boemare NE. *Steinernema cubana* sp. n. (Nematoda: Rhabditida: Steinernematidae) and the preliminary characterization of its associated bacterium. J. Inver. Path. 1994;64:123-129.

[32] Cabanillas HE, Poinar Jr GO, Raulston JR. *Steinernema riobravis* n. sp. (Rhabditida: Steinernematidae) from Texas. Fundam. Appl. Nematol. 1994;17:123-131.

[33] Nguyen KB, Buss EA. *Steinernema phyllophagae* n. sp. (Rhabditida: Steinernematidae), a new entomopathogenic nematode from Florida, USA. Nematol. 2011;13:425-442.

[34] Román J, Figueroa W. *Steinernema puertoricensis* n. sp. (Rhabditida: Steinernematidae), a new entomopathogenic nematode from Puerto Rico. J. Agril. Univ. Puerto. Rico. 1994;78:167-175.

[35] Qiu L, Zhao J, Wu Z, Lv Z, Pang Y. *Steinernema pui* sp. n. (Rhabditida, Steinernematidae), a new entomopathogenic nematode from Yunnan, China, Zootaxa. 2011;2767:3-11.

[36] Tallosi B, Peters A, Ehlers R-U. *Steinernema bicornutum* sp. n. (Rhabditida: Steinernematidae) from Vojvodina, Yugoslavia. Russ. J. Nematol. 1995;3:71-80.

[37] Ma J, Chen S, Clercq DEP, Han R, Moens M. *Steinernema changbaiense* sp. n. (Rhabditida: Steinernematidae), a new species of entomopathogenic nematode from Northeast China. Russ. J. Nematol. 2012a;20:97-12.

[38] Liu J, Berry RE. *Steinernema oregonensis* n. sp. (Rhabditida: Steinernematidae) from Oregon, USA. Fundam. Appl. Nematol. 1996;19:375-380.

[39] Tamirou T, Waeyenberg L, Tesfaye H, Ehlers R-U, Půža V, Mráček Z. *Steinernema ethiopiense* sp. n. (Rhabditida: Steinernematidae), a new entomopathogenic nematode from Ethiopia. Nematol.2012;14: 741-757.

[40] Elawad SA, Ahmad W, Reid A. *Steinernema abbasi* sp. n. (Nematoda: Steinernematidae) from the Sultanate of Oman. Fundam. Appl. Nematol. 1997;20:433-442

[41] Ma J, Shulong C, Li, X, Richou H, Hari B, Khatri-Chhetri, Patrick DC, Maurice M. A new entomopathogenic nematode, *steinernema tielingense* n. sp. (Rhabditida: Steinernematidae), from North China. Nematol. 2012;14(3):321-338.

[42] Artyukhovsky AK. *Neoaplectana arenaria* nov. sp. (Steinernematidae, Nematoda) inducing nematode disease in chafers of the Voronezh region. Trudy Voroneszhskogo Gosudarstvennogo Zapovednika. 1967;15:94-100.

[43] Ma J, Chen S, De Clercq P, Waeyenberge L, Han R, Moens M. A new entomopathogenic nematode, *Steinernema xinbinense* n. sp. (Nematoda: Steinernematidae), from North China. Nematol. 2012;14(6):723-739.

[44] Jian B, Reid AP, Hunt DJ. *Steinernema ceratophorum* n. sp. (Nematoda: Steinernematidae) a new entomopathogenic nematode from northeast China. Syst. Parasitol. 1997;37:115-125.

[45] Ngo Kanga F, Phap QT, Waeyenberge L, Spiridonov SE, Hauser S, Moens M. Two new species of *Steinernema* Travassos, 1927 from the humid forest of southern Cameroon. Russ. J. Nematol. 2012;20:15-26.

[46] Stock SP, Choo HY, Kaya HK. An entomopathogenic nematode, *Steinernema monticolum* sp. n. (Rhabditida: Steinernematidae) from Korea with a key to other species. Nematologica. 1997;43:15-29.

[47] Waturu CN, Hunt DJ, Reid AP. *Steinernema karii* sp. n. (Nematoda: Steinernematidae), a new

entomopathogenic nematode from Kenya. Int. J. Nematol. 1997;7:65-75.

[48] Nthenga I, Malan AP, Knoetze R, Tiedt LR, Berry S. *Steinernema sacchari* n. sp. (Rhabditida: Steinernematidae), a new entomopathogenic nematode from South Africa. Nematol. 2014; 16(4):475-494. doi:10.1163/15685411-00002780

[49] Stock SP, Somsook V, Reid A. *Steinernema siamkayai* n. sp. (Rhabditida: Steinernematidae), an entomopathogenic nematode from Thailand. Syst. Parasitol. 1998;41:105-113.

[50] Çimen H, Lee MM, Hatting J, Hazir S, Stock SP. *Steinernema tophus* sp. n. (Nematoda: Steinernematidae), a new entomopathogenic nematode from South Africa. Zootaxa. 2014; 3821:337-353.

[51] Luc PV, Nguyen KB, Reid AP, Spiridonov SE. *Steinernema tami* sp. n. (Rhabditida: Steinernematidae) from Cat Tein Forest, Veitnam. Russ. J. Nematol. 2000;8:33-43.

[52] Phan KL, Mrácek Z, Půža V, Nermuť J, Jarošová A. *Steinernema huense* sp. n., a new entomopathogenic nematode (Nematoda: Steinernematidae) from Vietnam. Nematol. 2014;16: 761-775.

[53] Phan KL, Nguyen NC, Moens M. *Steinernema loci* sp. n. and *Steinernema thanhi* sp. n. (Rhabditida: Steinernematidae) from Vietnam. Nematol. 2001;3:503-514.

[54] Mráček Z, Půža V, Nermut' J. *Steinernema poinari* sp. n. (Nematoda: Steinernematidae) a new entomopathogenic nematode from the Czech Republic. Zootaxa. 2014;3760:336-350.

[55] Phan KL, Nguyen NC, Moens M. *Steinernema sangi* sp. n. (Rhabditida: Steinernematidae) from Vietnam. Russ. J. Nematol. 2001;9:1-7.

[56] Çimen H, Lee MM, Hatting J, Hazir S, Stock SP. *Steinernema innovationi* n. sp. (Panagrolaimomorpha: Steinernematidae), a new entomopathogenic nematode species from South Africa. Helminthol. 2015;89:415-427.

[57] Malan AP, Knoetze R, Tiedt L. *Steinernema jeffreyense* n. sp. (Rhabditida: Heterorhabditidae), a new entomopathogenic nematode from South Africa. Helminthol. 2015;90:262-278.

[58] Shahina F, Anis M, Reid AP, Rowe J, Maqbool MA. *Steinernema pakistanense* sp. n. (Rhabditida: Steinernematidae) from Pakistan. Int. J. Nematol. 2001;11:124-133.

[59] San-Blas E, Portillo E, Nermut' J, Půža V, Morales-Montero P. *Steinernema papillatum* n. sp. (Rhabditida: Steinernematidae), a new entomopathogenic nematode from Venezuela. Nematol, 2015;17:1081-1097.

[60] Anis M, Shahina F, Reid AP, Rowe J. *Steinernema asiaticum* sp. n. (Rhabditida: Steinernematidae) from Pakistan. Int. J. Nematol. 2002;12:220-231.

[61] Hazir S, Faktorová, L, Çimen H, Nermuť J, Půža V Ramakuwela T, Hatting J. *Steinernema beitlechemi* n. sp., a new entomopathogenic nematode (Nematoda: Steinernematidae) from South Africa: Nematol. 2016;18(4):439-453.

[62] Nguyen KB, Duncan LW. *Steinernema diaprepesi* n. sp. (Rhabditida: Steinernematidae), a parasite of the citrus weevil *Diaprepes abbreviatus* (L.) (Coleoptera: Curculionidae). J. Nematol. 2002;34:159-170.

[63] Půža V, Nermut J, Mráček Z Gengler S, Haukeland S. *Steinernema pwaniensis* n. sp., a new

entomopathogenic nematode (Nematoda: Steinernematidae) from Tanzania. Helminthol. 2017;91(1):20-34. doi:10.1017/s0022149x15001157

[64] Hazir S, Stock SP, Keskin N. A new entomopathogenic nematode, *Steinernema anatoliense* n. sp. (Rhabditida: Steinernematidae), from Turkey. Syst. Parasitol. 2003;55:211-220.

[65] Abate BA, Malan AP, Tiedt LR, Wingfield MJ, Slippers B, Hurley BP. *Steinernema fabii* n. sp. (Rhabditida: Steinernematidae), a new entomopathogenic nematode from South Africa. Nematol. 2016;18:235-255.

[66] Stock SP, Koppenhöfer AM. *Steinernema scarabaei* n. sp. (Rhabditida: Steinernematidae), a natural pathogen of scarab beetle larvae (Coleoptera: Scarabaeidae) from New Jersey, USA. Nematol. 2003;5:191-204.

[67] Malan AP, Knoetze R, Tiedt L. *Steinernema nguyeni* n sp. (Rhabditida: Steinernematidae), a new entomopathogenic nematode from South Africa. Nematol. 2016;18:571-590.

[68] Mráček Z, Sturhan D, Reid A. *Steinernema weiseri* n. sp. (Rhabditida, Steinernematidae), a new entomopathogenic nematode from Europe. Syst. Parasitol. 2003;56:37-47.

[69] Cimen H, Půža V, Nermuť J, Hatting J, Ramakuwela T, Hazir S. *Steinernema biddulphi* n. sp., a new entomopathogenic nematode (Nematoda: Steinernematidae) from South Africa. J. Nematol. 2016;48:148-158.

[70] Triggiani O, Mráček Z, Reid A. *Steinernema apuliae* sp. n. (Rhabditida: Steinernematidae): a new entomopathogenic nematode from southern Italy. Zootaxa. 2004;460:1-12.

[71] Grifaldo-Alcantara PF, Alatorre-Rosas R, Segura-León O, Hernandez-Rosas F. *Steinernema ralatorei* n.sp. isolated from sugarcane areas at Veracruz, Mexico. Southwestern Entomologist. 2017;42:171-190.

[72] Qiu L, Fang Y, Zhou Y, Pang Y, Nguyen KB. *Steinernema guangdongense* sp. n. (Nematoda:Steinernematidae) a new entomopathogenic nematode from southern China with a note on *S. serratum* (nomen nudum). Zootaxa. 2004;704:1-20.

[73] Steyn WP, Knoetze R, Tiedt TR, Malan AP. *Steinernema litchii* n. sp. (Rhabditida: Steinernematidae), a new entomopathogenic nematode from South Africa. Nematol. 2017;19:1157-1177.

[74] Stock SP, Griffin CT, Chaerani R. Morphological and molecular characterization of *Steinernema hermaphroditum* n. sp. (Nematoda: Steinernematidae), an entomopathogenic nematode from Indonesia, and its phylogenetic relationships with other members of the genus. Nematol. 2004;6:401-412.

[75] Gorgadze O, Fanelli E, Lortkhipanidze M, Troccoli A, Burjanadze M, Tarasco E, Luca FD. *Steinernema borjomiense* n. sp. (Rhabditida: Steinernematidae), a new entomopathogenic nematode from Georgia. Nematol. 2018;pp. 1-17.

[76] Spiridonov SE, Krasomil-Osterfeld K, Moens M. *Steinernema jollieti* sp. n. (Rhabditida: Steinernematidae), a new entomopathogenic nematode from the American Midwest. Russ. J. Nematol. 2004;12:85-95.

[77] Stock SP, Campos-Herrera R, El-Borai FE, Duncan LW. *S. khuongi* n. sp. (Panagrolaimomorpha, Steinernematidae), a new entomopathogenic nematode species from Florida, USA. Helminthol, 2018;1-16. doi:10.1017/s0022149x18000081

[78] Yoshida M. *Steinernema litorale* n. sp. (Rhabditida:Steinernematidae), a new entomopathogenic nematode from Japan. Nematol. 2004;6:819-838.

[79] Tseng CT, Hou RF, Tang LC. *Steinernema taiwanensis* n. sp. (Rhabditida: Steinernematidae), a new entomopathogenic nematode from Taiwan. Zootaxa 2018;4434:466-480.

[80] Nguyen KB, Tesfamariam M, Gozel U, Gaugler R, Adams BJ. *Steinernema yirgalemense* n. sp. (Rhabditida: Steinernematidae) from Ethiopia. Nematol. 2004;6:839-856.

[81] Katumanyane A, Malan AP, Tiedt LR, Hurley BP. *Steinernema bertusi* n. sp. (Rhabditida: Steinernematidae), a new entomopathogenic nematode from South Africa. Nematol. 2020; 22:343-360.

[82] Qiu L, Yan Y, Zhou Y, Nguyen KB, Pang Y. *Steinernema aciari* sp n. (Nematoda: Steinernematidae), a new entomopathogenic nematode from Guangdong, China. J. Inver. Path. 2005a;88:58-69.

[83] Půža V, Campos-Herrera R, Blanco-Pérez R, Jakubíková H, Vicente-Díez I, Nermuť J. *Steinernema riojaense* n. sp., a new entomopathogenic nematode (Nematoda: Steinernematidae) from Spain. Nematol. 2020;22:825-844. https://doi.org/10.1163/15685411-00003343.

[84] Qiu L, Hu X, Zhou Y, Mei S, Nguyen KB, Pang Y. *Steinernema akhursti* n. sp. (Nematoda: Steinernematidae) from Yunnan, China. J. Inver. Path. 2005b;90:151-160.

[85] Lis M, Sajnaga E, Skowronek M, Wiater A, Rachwał K, Kazimierczak W. Steinernema sandneri n. sp. (Rhabditida: Steinernematidae), a new entomopathogenic nematode from Poland. J. of Nematol. 2021;53:1-24.

[86] Qiu L, Hu X, Zhou Y, Pang Y, Nguyen, KB. *Steinernema beddingi* n. sp. (Nematoda: Steinernematidae), a new entomopathogenic nematode from Yunnan, China. Nematol. 2005;7:737-749.

[87] Poinar GO Jr. Description and biology of a new insect parasitic rhabitoid, *Heterorhabditis bacteriophora* n. gen. n. sp. (Rhabditida; Heterorhabditidae n. family). Nematologica, 1976;21: 463-470.

[88] Phan KL, Subbotin SA, Waeyenberge L, Moens M. A new entomopathogenic nematode, *Steinernema robustispiculum* n. sp. (Rhabditida: Steinernematidae) from Chumomray National Park in Vietnam. Syst. Parasitol. 2005;60:23-32.

[89] Poinar GO Jr, Jackson T, Klein M. *Heterorhabditis megidis* sp. n. (Heterorhabditidae: Rhabditidia), parasitic in Japanese beetle, *Popillia japonica* (Scarabidae: Coleoptera), in Ohio. Proc. Helminthol. Soc. Wash. ISSN: 0018-0130 1987; 53:53-59.

[90] Sturhan D, Spiridonov SE, Mráček Z. *Steinernema silvaticum* sp. n. (Rhabditida: Steinernematidae), a new entomopathogenic nematode from Europe. Nematol. 2005;7: 227-241.

[91] Poinar GO Jr. Taxonomy and biology of Steinernematidae and Heterorhabitidae. In: Enromopathogenic Nemartodes in Biological Control (Gaugler R and Kaya HK, eds.). Boca Raton, FL, USA, CRC Press. 1990. pp. 23-61.

[92] Phan LK, Takemoto S, Futai K. *Steinernema ashiuense* sp. n. (Nematoda: Steinernematidae), a new entomopathogenic nematode from Japan. Nematol. 2006b;8: 681-690.

[93] Poinar GO Jr, Karunakar GK, David H. *Heterorhabditis indicus* n. sp. (Rhabditida, Nematoda) from India: separation of *Heterorhabditis* spp. by infective juveniles. Fundam. Appl. Nematol. 1992;15:467-472.

[94] Phan KL, Spiridonov SE, Subbotin SA, Moens M. Four new species of *Steinernema* Travassos, 1927 with short infective juvenile from Vietnam. Russ. J. Nematol. 2006;14:11-29.

[95] Liu J, Berry RE. *Heterorhabditis marelatus* n. sp. (Rhabditida:Heterorhabditidae) from Oregon. J. Invertebr. Pathol. 1996;67:48-54.

[96] Shamseldean MM, Abou El-Sooud AB, Abd-Elgawad, MM, Saleh, MM. Identification of a new *Heterorhabditis* species from Egypt, *Heterorhabditis taysearae* n. sp. (Rhabditida: Heterorhabditidae). Egypt. J. Biol. Pest Co.1996;6:129-138.

[97] Stock SP, Griffin CT, Burnell AM. Morphological characterization of three isolates of *Heterorhabditis* Poinar, 1976 from the 'Irish group' (Nematoda: Rhabditida: Heterorhabditidae) and additional evidence supporting their recognition as a distinct species, *H. downesi* n. sp. Syst. Parasitol. 2002;51:95-106.

[98] Phan KL, Subbotin SA, Nguyen NC, Moens M. *Heterorhabditis baujardi* sp. n. (Rhabditida: Heterorhabditidae) from Vietnam and morphometric data for *H. indica* populations. Nematol, 2003;5: 367-382.

[99] Nguyen KB, Malan AP, Gozel U. *Steinernema khoisanaen*. sp. (Rhabditida: Steinernematidae), anew entomopathogenic nematode from South Africa. Nematol. 2006;8:157-175.

[100] Nguyen K, James R, McCoy C, Adams B Stuart R, Shapiro-Ilan D. *Heterorhabditis mexicana* n. sp. (Rhabditida: Heterorhabditidae) from Tamaulipas, Mexico, and morphological studies of the bursa of *Heterorhabditis* spp. Nematol. 2004;6(2): 231-244. doi:10.1163/1568541041218031

[101] Nguyen KB, Qiu L, Zhou Y, Pang Y. *Steinernema leizhouense* sp. n. (Nematoda: Steinernematidae), a new entomopathogenic nematode from southern China. Russ. J. Nematol. 2006;14(2):101-118.

[102] Andaló V, Nguyen KB, Moino A. *Heterorhabditis amazonensis* n. sp. (Rhabditida: Heterorhabditidae) from Amazonas, Brazil. Nematol. 2006;8:853-867.

[103] Mráček Z, Nguyen KB, Tailliez P, Boemare N, Chen S. *Steinernema sichuanense* n. sp. (Rhabditida, Steinernematidae), a new species of entomopathogenic nematode from the province of Sichuan, east Tibetan Mts., China. J. Invertebr. Pathol. 2006;93(3):0-169. doi:10.1016/j.jip.2006.06.007

[104] Nguyen KB. Gozel U, Koppenhöfer H, Byron J, Adams BJ. *Heterorhabditis floridensis* n. sp. (Rhabditida: Heterorhabditidae) from Florida. Zootaxa, 1177(1):21.

[105] Chen S, Yan A, Li X, Moens M, Spiridonov S. A new entomopathogenic nematode, *Steinernema hebeiense* sp. n. (Rhabditida: Steinernematidae), from North China. Nematol. 2006;8:563-574.

[106] Nguyen KB, Shapiro-Ilan D, Mbata G. *Heterorhabditis georgiana* n. sp. (Rhabditida: Heterorhabditidae) from Georgia, USA. Nematol. 2008;10(3):433-448. doi:10.1163/156854108783900276

[107] Uribe-Lorío L, Mora M, Stock SP. *Steinernema costaricense* n. sp. and *Steinernema puntauvense* n. sp. (Rhabditida: Steinernematidae), two

new entomopathogenic nematodes from Costa Rica. Syst. Parasitol. 2007;68:167-182.

[108] Malan AP, Nguyen KB, De Waal JY, Tiedt L. *Heterorhabditis safricana* n. sp. (Nematoda: Heterorhabditidae), a new entomopathogenic nematode from South Africa. Nematol. 2008;10:381-396.

[109] Edgington S, Buddie AG, Moore D, France A, Merino L, Hunt DJ. *Heterorhabditis atacamensis* n. sp. (Nematoda: Heterorhabditidae), a new entomopathogenic nematode from the Atacama Desert, Chile. J. Helminthol. 2011;85:381-394.

[110] Nguyen KB, Hunt DJ. Entomopathogenic nematodes: Systematics, phylogeny and bacterial symbionts. Nematology Monographs and Perspectives. *Brill,* Leiden-Boston, the Netherlands, 2007;5:816.

[111] Xing-Yue LI. Qi-Zhi L, Nermuť J, Půža, V, Mráček Z. *Heterorhabditis beicherriana* n. sp. (Nematoda: Heterorhabditidae), a new entomopathogenic nematode from the Shunyi district of Beijing, China. Zootaxa. 2012;3569(1):25-42. doi:10.11646/zootaxa.3569.1.2

[112] Nguyen KB, Půža V, Mráček Z. *Steinernema cholashanense* n. sp. (Rhabditida, Steinernematidae) a new species of entomopathogenic nematode from the province of Sichuan, Chola Shan Mountains, China. J. Invertebr. Pathol. 2008;97:251-264.

[113] Malan AP, Knoetze R, Tiedt L. *Heterorhabditis noenieputensis* n. sp. (Rhabditida: Heterorhabditidae), a new entomopathogenic nematode from South Africa. J. Helminthol. 2014;88:139-151.

[114] Gaugler R, Kaya HK. Entomopathogenic Nematodes in Biological Control. Boca Raton, FL, USA, CRC Press. 1990. ISBN-0849345413.

[115] Gaugler R, Han R. Production technology. *In:* Entomopathogenic Nematology (Gaugler R. eds.) Wallingford, UK: CABI. 2002. pp. 289-310

[116] Akhurst RJ, Boemare NE. A numerical taxonomic study of the genus *Xenorhabdus* (Enterobacteriaceae) and proposed elevation of the subspecies of *X. nematophilus* to species. J. Gen. Microbiol. 1988;134:1835-1845.

[117] Griffin CT, Downes MJ, Block W. Tests of Antarctic soils for insect parasitic nematodes. Antarct. Sci. 1990;2(03). doi:10.1017/s095410209000030x

[118] Kaya HK. Soil ecology. In: Entomopathogenic Nematodes: *Biological Control* (Gaugler R, Kaya HK. eds.). CRC Press, Boca Raton, Florida, 1990.pp. 93-115.

[119] Mráček Z, Weiser J, Gerdin S. Head and cuticular structures of some species in the family Steinernematidae (Nematoda). Nematologica. 1981;27:443-448.

[120] Selvan SP, Grewal PS, Gaugler R, Tomalak M. Evaluation of steinernematid nematodes against *Popillia japonica*: species, strain and rinse after application. J. Econ. Entomol. 1994;89:605-609.

[121] Campbell J, Gaugler R. Nictation behavior and its ecological implications in the host search strategies of entomopathogenic nematodes (Heterorhabditidae and Steinernematidae). Behaviour. 1993;126:154-169.

[122] Nguyen KB, Smart Jr GC. *Steinernema scapterisci* n. sp. (Steinernematidae: Nematoda). J. Nematol. 1990;22:187-199.

[123] Grewal, PS, Nardo EABDE, Aguillera MM. Entomopathogenic

nematodes: potential for exploration and use in South America. Neotrop. Entomol. 2001; 30(2):191-205.

[124] Bedding RA, Molyneux AS. Penetration of insect cuticle by infective juveniles of *Heterorhabditis* spp. (Heterorhabditidae: Nematoda). Nematologica. 1982;28:354-359.

[125] Hill D E. Entomopathogenic nematodes as control agents of developmental stages of the black-legged tick, *Ixodes scapularis*. J. Parasitol. 1998;84(6):pp. 1124. doi:10.2307/3284660

[126] Johnigk SA, Ehlers RU. Juvenile development and life cycle of *Heterorhabditis bacteriophora* and *H. indica* (Nematoda: Heterorhabditidae). Nematol. 1999a;1:251-260.

[127] Griffin CT, O'callaghan KM, Dix I. A self-fertile species of *Steinernema* from Indonesia: further evidence of convergent evolution amongst entomopathogenic nematodes? Parasitol., 2001;122(02). doi:10.1017/s003118200100717x

[128] Hominick WM, Reid AP, Bohan DA, Briscoe BR. Entomopathogenic nematodes: biodiversity, geographical distribution and the convention on biological diversity. Biocon. Sci. Technol. 1996;6:317-331.

[129] Zograf J, Borgonie G, Bert W. The structure of the female reproductive system of nematodes from the genus *Steinernema* (Rhabditida: Steinernematidae). Nematol. 2008;10(6): 883-896. doi:10.1163/156854108786161463

[130] Strauch O, Stoessel S, Ehlers RU. Culture conditions define automictic or amphimictic reproduction in entomopathogenic rhabditid nematodes of the genus *Heterorhabditis*. Fundam. Appl. Nematol. 1994;17:575-582.

[131] Lewis EE. Behavioral Ecology. In: Entomopathogenic Nematology (Gauger R. ed.). New York, CAB International, 2002. pp. 205-223.

[132] Ebssa L, Dix I, Griffin CT. Female presence is required for male sexual maturity in the nematode Steinernema longicaudum. Curr. Biol., 2008;18(21):R997–R998. doi:10.1016/j.cub.2008.09.032

[133] Ebssa L, Borgemeister C, Poehling HM. Effectiveness of different species/strains of entomopathogenic nematodes for control of western flower thrips (*Frankliniella occidentalis*) at various concentrations, host densities, and temperatures. Biol. Cont. 2004;29(1):145-154. doi:10.1016/s1049-9644(03)00132-4

[134] Schroeder WJ. Laboratory bioassays and field trials of entomogenous nematodes for control of *Diaprepes abbreviatus* (Coleoptera: Curculionidae) in Citrus. Environ. Entomol. 1987;16(4):987-989. doi:10.1093/ee/16.4.987

[135] Bastidas B, Edgar P, San-Blas E. Size does matter: The life cycle of *Steinernema* spp. in micro-insect hosts. J. Invertebr. Pathol. 2014;121:46-55.

[136] Lewis EE, Campbell J, Griffin C, Kaya H, Peters A. Behavioral ecology of entomopathogenic nematodes. Biol. Control. 2006;38:66-79.

[137] Dillon A. *Biological control of the large pine weevil, Hylobius abietis L., (Coleoptera: Curculionidae) using entomopathogenic nematodes*. Dissertation. National University of Ireland–Maynooth, Maynooth, Ireland. 2003.

[138] Campbell JF, Kaya HK. Variation in entomopathogenic nematode (Steinernematidae and Heterorhabditidae) infective-stage jumping behaviour. Nematol. 2002;4:471-482.

[139] Reed EM, Wallace HR. Leaping locomotion in an insect parasitic nematode. Nat. 1965;206:210-211.

[140] Campbell JF, Kaya HK. How and why a parasitic nematode jumps. Nat. 1999a;397:485-486.

[141] Campbell JF, Kaya HK. Mechanism, kinematic performance, and fitness consequences of entomopathogenic nematode (*Steinernema* spp.) jumping behavior. Can. J. Zool. 1999b;77:1947-1955.

[142] Kruitbos LM, Heritage S, Hapca S, Wilson MJ. The influence of habitat quality on the foraging strategies of the entomopathogenic nematodes *Steinernema carpocapsae* and *Heterorhabditis megidis*. Parasitol. 2010;137:303-309.

[143] Grewal PS, Selvan S, Gaugler R. Thermal adaptation of entomopathogenic nematodes – niche breadth for infection, establishment and reproduction. J. Therm. Biol. 1994;19:245-253.

[144] Lewis EE, Gaugler R, Harrison R. Entomopathogenic nematode host finding: Response to host contact cues by cruise and ambush foragers. Parasitol. 1992;105:309-315.

[145] Lewis EE, Grewal PS Gaugler R. Hierarchical order of host cues in parasite foraging: A question of context. Parasitol. 1995;110:207-213.

[146] Koppenhöfer AM, Grewal PS, Fuzy EM. Differences in penetration routes and establishment rates of four entomopathogenic nematode species into four white grub species. J. Invertebr. Pathol. 2007;35:128-139.

[147] Adams BJ, Fodor A, Koppenhöfer HS, Stackenbrandt E, Stock SP, Klein MG. Biodiversity and systematic of nematode–bacterium entomopathogens. Biol. Control. 2006;38:4-21.

[148] Shairra SA. Paeasitizm of locust by entomopathogenic nematode in relation to insect micro-aggregation inhibitor. Egypt. Acad. J. Biol. Sci. 2009;2(2):221-230.

[149] Kaya HK, Gaugler R. Entomopathogenic Nematodes. Ann. Rev. Entomol. 1993;38:181-206.

[150] Sandhu SK, Jagdale GB, Hogenhout SA, Grewal PS. Comparative analysis of the expressed genome of the infective juvenile entomopathogenic nematode, *Heterorhabditis bacteriophora*. Mol. Biochem. Parasitol. 2006;145: 239-244.

[151] Shapiro-Ilan DI, Han R, Dolinksi C. Entomopathogenic nematode production and application technology. J. Nematol. 2012;44:206-217.

[152] Grewal PS, Ehlers RU, Shapiro-Ilan DI. Nematodes as biological control agents. Wallingford: CABI Publishing. 2005. ISBN- 0851990177.

[153] Hominick WM, Briscoe BR, del Pino FG, Heng J, Hunt DJ, Kozodoy E, Mracek Z, Nguyen KB, Reid AP, Spiridonov S, Stock SP, Sturhan D, Waturu C, Yoshida, M. Biosystematics of entomopathogenic nematodes: current status, protocols and definitions. J. Helminthol. 1997;71(04):271.

[154] Poinar, GO. Origins and phylogenetic relationships of the entomophilic rhabditis, *Heterorhabditis* and *Steinernema*. Fundam. Appl. Nematol. 1993;16(4):333-338.

[155] Adams B, Nguyen KB. Taxonomy and systematics. In: Entomopathogenic Nematology (Gaugler R ed.). CABI Publishing, Wallingford, UK, 2002. pp. 1-33.

[156] Nguyen KB, Smart GC Jr. Identification of entomopathogenic nematodes in the Steinernematidae and

Heterorhabditidae (Nemata: Rhabditida). J. Nematol. 1996;28:286-300.

[157] Nguyen KB, Smart GC Jr. Scanning electron microscopic studies of spicules and gubernacula for *Steinernema* spp. (Nemata: Steinernematidae). Nematologica, 1997;43:465-480.

[158] Spiridonov SE, Hominick WM, Brisco BR. Morphology of amoeboid cells in the uterus of *Steinernema* species (Rhabditida: Steinernematidae). Russ. J. Nematol. 1999;7:39-42.

[159] Mráček Z, Bednarek A. The morphology of lateral fields of infective juveniles of entomogenous nematodes of the family Steinernematidae (Rhabditida). Nematologica. 1991;37:63-71.

[160] Jackson GJ. Differentiation of three species of *Neoaplectana* (Nemtoda:Rhabditidia) grown axenically. Parasitol. 1965;55:571-578.

[161] Akhurst RJ. Use of starch gel electrophoresis in the taxonomy of the genus *Heterorhabditis* (Nematoda: Heterorhabditidae). Nematologica. 1987;33:1-9.

[162] Curran J, Webster JM. Genotypic analysis of *Heterorhabditis* isolates from North Carolina. J. Nematol. 1989;21:140-145.

[163] Smits PH, Groenen TM, De Raay G. Characterization of *Heterorhabditis* isolates using DNA restriction fragment 1ength polymorphism. Revue de Nématologie. 1991;14:445-453.

[164] Reid AP, Hominick WH. Cloning of the rDNA repeat unit from a British entomopathogenic nematode (Steinernematidae) and its potential for species identification. Parasitol. 1993;107:529-536.

[165] Vrain TC, Wakarchuk DA, Levesque AC, Hamilton RI. Intraspecific rDNA restriction fragment length polymorphisms in the *Xiphinema americanum* group. Fundam. Appl. Nematol.1992;15:563-574.

[166] Curran J, Driver F. Molecular taxonomy of *Heterorhabditis*. Cost 812 Biotechnology: genetics of entomopathogenic nematode-bacterium complexes. In: Proceedings of symposium and workshop, St Patrick's College, Maynooth, Kildare County, Ireland. Luxembourg, European Commission, Dgeur, (Burnell AM, Ehlers RU, Masson JP. Eds.). 1994;15681:41-48.

[167] Joyce SA, Burnell AM, Powers TO. Characterization of *Heterorhabditis* isolates by PCR amplification of segments of mtDNA and rDNA genes. J. Nematol. 1994;26:260-270.

[168] Reid AP. Molecular taxonomy of Stcille/ïlcma. In: Proceedings of a symposium and workshop, St Patrick's College, Maynooth, Co. Kildare, Ireland, E.e. DG XII, Luxembourg (Bumell AM, Ehlers RU and Masson JP. Eds.) CaST 812 Biotechnology: *Genclics of entomopathogellicnernatode-bacterillrn complexes*. 1994pp. 49-58.

[169] Subbotin SA, Moens M. Molecular taxonomy and phylogeny. In: Plant nematology (Perry RN, Moens M. eds.). Wallingford, UK, CABI Publishing, 2006. pp. 33-58.

[170] Didiza L, Lephoto TE , Gray VM. Morphological and molecular phylogenetic description of Steinernema batswanae n. sp. (Rhabditida: Steinernematidae): a new species of an entomopathogenic nematode from South Africa. Arch Phytopathol Plant Prot. 2021 DOI: 10.1080/03235408.2021.1931648

[171] Bhat AH, Chaubey AK, Hartmann J, Nermut J, Puza V. Notes on the morphology, bionomics, distribution, and efficacy of Steinernema siamkayai (Rhabditida:

Steinernematidae) from western Uttar Pradesh, India. Nematol. 2021; 54:817-836. DOI: 10.1163/15685411-bja10079

[172] Powers TO, Todd TC, Burnell AM, Munay PCB, Fleming CC, Szalanski AL, Adams BA, Harris TS. The rDNA internal transcribed spacer region as a taxonomic marker for nematodes. J. Nematol., 1997;29:441-450.

[173] Blaxter ML, De Ley P, Garey JR, Liu LX, Scheldeman P, Vierstraete A, Vanfleteren JR, Mackey LY, Dorris M, Frisse LM, Vida JT, Thomas WK. A molecular evolutionary framework for the phylum Nematoda. Nat. 1998;392:71-75.

[174] Dorris M, De Ley P, Blaxter ML. Molecular analysis of nematode diversity and the evolution of parasitism. Parasitol. Today 1999;15:188-193.

[175] Liu J, Berry R, Poinar G, Moldenke A. Phylogeny of *Photorhabdus* and *Xenorhabdus* species and strains as determined by comparison of partial 16S rRNA gene sequences. Int. J. Syst. Bacteriol. 1997;47:948-951.

[176] Adams BJ, Burnell AM, Powers TO. A phylogenetic analysis of *Heterorhabditis* (Nematoda:Rhabditidae) based on internal transcribed spacer 1 DNA sequence data. J. Nematol. 1998;30:22-39.

[177] Szalanski AL, Taylor DB, Mullin PG. Assessing nuclear and mitochondrial DNA sequence variation within *Steinernema* (Rhabditida: Steinernematidae). J. Nematol. 2000;32:229-233.

[178] Stock SP, Campbell JF, Nadler SA. Phylogeny of *Steinernema* Travassos, 1927 (Cephalobina: Steinernematidae) inferred from ribosomal DNA sequences and morphological characters. J. Parasitol. 2001;87:877-889.

[179] Bhat AH, Istkhar R, Chaubey AK, Půža V, San-Blas E. First Report and Comparative Study of Steinernema surkhetense (Rhabditida: Steinernematidae) and its Symbiont Bacteria from Subcontinental India. J Nematol. 2017; 49(1):92-102. doi: 10.21307/jofnem-2017-049.

[180] Hall R. Challenges and prospects of integrated pest management. In: Novel approaches to integrated pest management (Reuveni R. ed.). Lewis Publishers, Boca Raton, Florida, USA. 1995.p. 1-19.

[181] Dhaliwal GS, Dhawan AK, Singh R. Biodiversity and ecological agriculture: Issues and perspectives. Indian J. Ecol. 2007;34(2):100-109.

[182] Pimentel D. Area-wide pest management: Environmental, economic and food issues. In: Area-wide control of insect pests: from research to field implementation (Vreysen MJB, Robinson AS and Hendrichs J. eds.). Springer, Dordrecht, the Netherlands, 2007. pp. 35-47.

[183] Pimentel D. Pesticides and pest control. In: Integrated pest management innovation – development (Peshin R, Dhawan AK. eds.). Springer, Dordrecht, the Netherlands, 2009.pp. 83-88.

[184] Caltagirone LE, Doutt RL. The history of the vedalia beetle importation to California and its impact on the development of biological control. Annu. Rev. Entomol. 1989;34:1-16.

[185] Ferron P. Biological control of insect pests by entomogenous fungi. Annu. Rev. Entomol. 1978;23: 409-442.

[186] Priest F. Biological control of mosquitoes and other biting flies *by Bacillus sphaericus* and *Bacillus thuringiensis*. J. Appl. Microbiol. 1992;72: 357-369.

[187] Georgis R, Koppenhöfer AM, Lacey LA, Bélair G, Duncan LW, Grewal PS, Samish M, Tan L, Torr P, van Tol RWHM. Successes and failures in the use of parasitic nematodes for pest control. Biol. Control. 2006;38:103-123.

[188] Bhat AH, Rana A, Chaubey AK, Shokoohi E, Machado RAR. Characterisation of Steinernema abbasi (Rhabditida: Steinernematidae) isolated from Indian agricultural soils and their efficacy against insect pests. Biocontrol Sci Technol. 2021; 30:8 DOI: 10.1080/09583157.2021.1917514

[189] Malan AP, Manrakhan A. Susceptibility of the Mediterranean fruit fly (*Ceratitis capitata*) and the Natal fruit fly (*Ceratitis rosa*) to entomopathogenic nematodes. J. Invertebr. Pathol. 2009;100(1):47-49.

[190] Van Niekerk S, Malan AP. Potential of South African entomopathogenic nematodes (Heterorhabditidae and Steinernematidae) for control of the citrus mealy bug, *Planococcus citri* (Pseudococcidae). J. Invertebr. Pathol. 2012;111:166-176.

[191] Van Niekerk S, Malan AP. Adjuvants to improve control of *Planococcus citri* (Hemiptera: Pseudococcidae) using entomopathogenic nematodes (Heterorhabditidae and Steinernematidae). J. Helminthol. 2013. Doi: 19.1163/15685411-00002780.

[192] Van Niekerk S, Malan AP. Compatibility of *Heterorhabditis zealandica* and *Steinernema yirgalemense* with agrochemicals and biological control agents. Afr. Entomol. 2014a;22:49-56.

[193] Ganguly S. Taxonomy of entomopathogenic nematodes. In: *National Congress on Centenary of Nematology in India, Appraisal and Future Plans*. Division of Nematology, IARI, New Delhi, Abstract: 2001. pp. 20-21.

[194] Hunt DJ. Overview of taxonomy and systematics. In: *Entomopathogenic nematodes*: systematics, phylogeny and bacterial symbionts. Nematology Monographs and Perspectives (Nguyen KB, Hunt DJ. eds.). Leiden, the Netherlands: Brill Publishing; 2007;pp. 27-57.

[195] Stock SP, Griffin CT, Burnell AM. Morphological characterization of three isolates of *Heterorhabditis* Poinar, 1976 from the 'Irish group' (Nematoda: Rhabditida: Heterorhabditidae) and additional evidence supporting their recognition as a distinct species, *H. downesi* n. sp. Syst. Parasitol. 2002;51:95-106.

[196] Nadler SA, Bolotin E, Stock SP. Phylogenetic relationships of *Steinernema* Travassos, 1927 (Nematoda: Cephalobina: Steinernematidae) based on nuclear, mitochondrial and morphological data. Syst. Parasitol. 2006;63:161-181.

Chapter 13

New Approach for the Evaluation of Ecological Quality in the Mediterranean Coastal Ecosystems, Case Study of Bizerte Lagoon: Marine Nematodes Functional Traits Assessment

Ahmed Nasri, Patricia Aïssa, Hamouda Beyrem and Ezzeddine Mahmoudi

Abstract

Marine ecosystems have great economic and ecological value, as they provide good services and habitats for a variety of organisms. However, the marine environment is under anthropogenic stressors. The Mediterranean basin is one of the most threatened ecosystems, where urban and industrial waste is becoming a growing risk for coastal marine habitats integrity. The Bizerte lagoon represents a major coastal lagoon and is an example of such an aquatic environment continuously exposed to pollutants. Marine nematodes are the most diverse metazoans and represent an excellent model for the environmental monitoring because they can be easily sampled and maintained under experimental conditions. Nematode communities are investigated for the analysis of taxonomic diversity and ecological indices. Currently, we present here to evaluate the ecological quality based on the description of nematode assemblages using biological traits and functional groups. This relatively new approach allows obtaining insight into the status of marine coastal ecosystems.

Keywords: Mediterranean basin, Nematodes, Ecological quality, Chemicals pollutants, Ecological indices, Biological traits

1. Introduction

Aquatic environments and in particular coastal ecosystems are exposed to a variety of contaminants derived from human activities. The majority of pollutants is removed from the water column and accumulated in marine sediment. Among these chemical compounds, we can list pesticides commonly used worldwide as pest control agents in agriculture [1], persistent organic pollutants (POPs) such as polycyclic aromatic hydrocarbons (PAHs), polychlorinated biphenyls (PCBs), and dioxins realized from industrial sources and maritime traffic [2], heavy metals originating from discharges from the metallurgical industry [3], for human health applications [4], and microplastic particles by the direct release of textile fibers or by cleaning products [5, 6]. All these compounds can enter marine ecosystems through a variety of routes, including urban (parking lots and residential areas) and agricultural (treated agricultural areas) runoff, washout or spray drift, or of contaminated sediments. Once these pollutants arrive in aquatic ecosystems, they can persist for a few months to several years [7, 8].

Like most coastal areas of the Mediterranean Sea, the Bizerte lagoon in northern Tunisia is a coastal ecosystem exposed to pollutants resulting from agriculture, urbanization, industrialization, as well as pressures from maritime and commercial ports. Marine sediments are therefore highly contaminated with a wide range of chemicals. Thereby, free marine nematodes are considered a useful tool to assess the presence of impact in marine ecosystems since they spent their life in the benthic ecosystems inhabiting the first ten centimeters of marine sediments [9] therefore directly exposed to pollutants. Nematodes are ubiquitous (the abundance ranges from 11 to 24 million individuals per square meter), occupy a key role in the benthic food webs [10], play an important role in the ecosystems functioning [11] through sediment aeration [12], and the mineralization of organic matter [13]. Additionally, due to their short life cycles, rapid metabolic rates, benthic larval stages, and rapid responses to environmental changes, nematodes are considered an ideal model for laboratory experiments [14, 15], and are classified excellent indicators for biomonitoring activities [9].

This chapter aims to describe the results of previous studies using traditional monitoring tools (analysis of the taxonomic structure of nematodes and determination of ecological indices) to assess the environmental quality and to describe the effectiveness of the new multivariate analysis approach (creation of functional groups of nematodes bases on biological traits) in order to provide clearer and more informative data on the state of the Mediterranean coastal ecosystem.

2. Materials and methods

2.1 Sampling site

Sediments samples are collected in the upper layer from Bizerte lagoon (NE, Tunisia) using Plexiglas hand-cores and placed in a bucket (**Figure 1**). Only the first 10 cm were sampled like that the presence of 90% of the nematode is located from 1 to 2 cm from the surface [16].

2.2 Sediment contamination and nematodes study

Sediment samples used for the experiment of chemical enrichment were alternately frozen (−20°C) and thawed three times in order to eliminate all fauna according to [17] before adding the selected concentrations of contaminant used [18]. Particles larger than 63 μm were removed by wet sieving and selected concentrations of chemical compounds were added to 100 g of Dry weight (DW) sediment. 2-l-glass bottles were used as microcosms in the experiment [17]. The control microcosm (C) consisted of non-treated sediments containing 200 g of natural sediments and 100 g of defaunated sediments in 1 L of filtered water (1 μm filtration).

The treated microcosms were constituted by 300 g of homogenized sediments (two-thirds of natural sediments and one-third of contaminated sediments) in 1 L of filtered water as reported in [19]. Overall, one control [(C)] and n^{th}-enriched

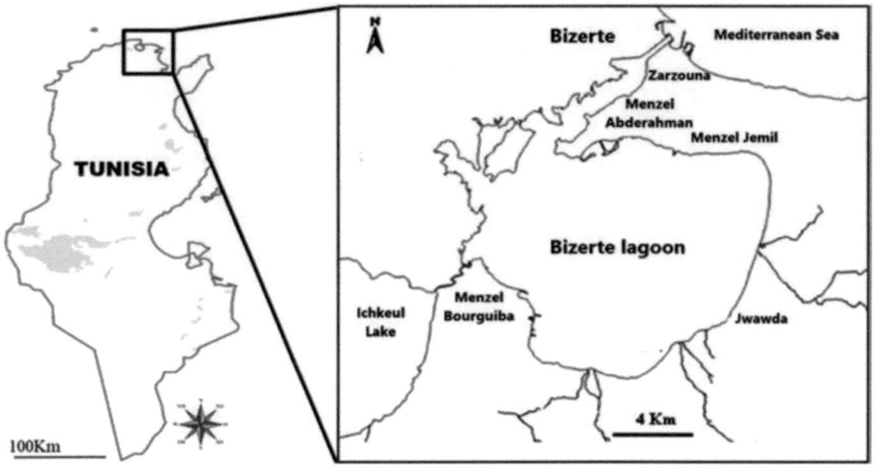

Figure 1.
Location of the Bizerte lagoon (northern Tunisia).

microcosms with contaminants were investigated [20]. Each treatment was replicated three times and the experiment was run for one month as reported by [21].

After generally 30 days, the experiment was conducted adding 4% buffered formalin solution in the experimental plots. Using the Ludox™ centrifugation technique and sieves (40 µm - 1 mm), the meiobenthic nematodes were extracted from the sediments and stained for 48 h with Rose Bengal [22]. Afterward, meiofaunal taxa were identified and counted under a stereomicroscope (50×, Wild Heerbrugg M5A Model), and a maximum of 100 nematodes/replicates were randomly collected. All specimens were placed in twenty percent of glycerol, evaporated to anhydrous glycerol, and mounted on slides [23]. The genus-level were identified according to the literature available [24] and NEMYS repository [25] using a Nikon microscope (Image Software NIS Elements Analysis Version 4.0 Nikon 4.00.07–build 787–64 bit).

2.3 Ecological indices analysis

The ecological diversity indices present tools that take into account both the number of species present and their relative abundance. In order to assess the effect of chemical pollutants on the benthic fauna, the former research work has focused on studying the univariate biodiversity indices of meiofaunal taxa as well as monitoring changes to the composition. The spatial or temporal diversity of a taxonomic group (Shannon Diversity), the distribution of the relative abundances of species (Equitability), the specific richness (the number of species present), Index of Maturity, and Trophic Diversity were studied. **Table 1** shows these indices in detail.

2.4 Functional traits analysis

Currently, the new research is based on the classification of nematodes into functional groups according to the diversity of their morphological attributes and life-history strategies that are related to ecological functions [31]. The combined analysis of functional traits was suggested for the first time by [32] in order to have more effective indications of the state of the environment. The assessment of functional diversity is based on the grouping of functionality measures into one using a multivariate approach that provides more informative ecological data than single

Index	Formula	Using	Reference
Shannon Diversity	$H' = -\Sigma$ (pi ln pi)	Most commonly used, makes it possible to assess the spatial or temporal diversity of specific taxonomic group.	[26]
Equitability	$J' = H'/ \ln S$	Equitability (J') provides details on the species/ genus/taxon relative contribution to the overall diversity.	[27]
Margalef Diversity	$D = (S-1)/\ln(N)$	D is the raw number of species in a sample. It takes into account both the number of species and the total sample size.	[28]
Index of Maturity	$MI = \Sigma^n i = 1$ $(vi \times fi)$	MI is considered as a measure of the assemblage's life strategy.	[29]
Trophic Diversity	$TD = \Sigma\theta 2$	High TD values indicate the dominance of a single trophic guild compared the overall assemblages.	[30]

Table 1.
The ecological indices used in marine nematodes.

analyzes. It is about making more reliable correlative relationships with environmental factors that can reveal additive relationships within communities.

In the **Table 2**, Five functional attributes were proposed for nematodes: (a) Adult length (1–2 mm, 2–4 mm, and ˃ 4 mm) [32]; Amphid shape (indistinct (Id); slit-like (St), blister-like (Bs); longitudinal slit (Ls); spiral (SP); rounded or elongate loop (REL); pocket-like (Pk), and circular (Cr) [32]; Tail shape (short/ round (s/r), elongated/filiform (e/f), conical (co), clavate/ conical–cylindrical (cla)) [33]; Life history (c-p scores; from 1 to 5) evaluated through a scale from colonizers to persistent [29, 34]; and Feeding diet: (selective deposit feeders (1A), non-selective deposit feeders (1B), epigrowth-feeders (2A), and omnivores-carnivores (2B)) [35].

2.5 Statistical data analysis

Nematodes data were tested for normality and homogeneity of variance using Kolmogorov–Smirnov, and Bartlett tests, respectively. The software PRIMER v5 (Plymouth Routines in Multivariate Ecological Research, version 5.1.8) will be used following the standard methods described by [36] for univariate and multivariate analysis of data [37]. For each condition, the univariate indices used: species numbers (S), species richness (d), and the Shannon diversity (H') and Pielou's evenness (J') indices. Subsequently, means of all univariate indices examined using a one-way analysis of variance (1-ANOVA) among all microcosms. Multiple comparisons were performed using the Tukey HSD test (software Statistica version 5.1). Significant differences were considered when p-values were < 0.05.

For multivariate analysis of nematode communities, the non-metric Multi-Dimensional Scaling (nMDS) ordination measures applied on square-root transformed functional traits abundances and using Bray–Curtis similarity to represent how treatments or biological attributes are matched. The SIMPER analysis run to determine the contribution of each genus or functional group (cumulative contribution of 70%) to the average dissimilarity among treatments.

3. Results and discussion

Ecotoxicological studies using the traditional approach for aquatic ecosystem monitoring considered nematodes diversity and taxonomic composition. The most

DOI: http://dx.doi.org/10.5772/intechopen.98815

Genera	Functional traits				
	1A, 1B, 2A and 2B	C-p score	Amphid shape	Tail shape	Adult length
Daptonema	1B	2	Cr	cla	1–2 mm
Metalinhomoeus	1B	2	Cr	e/f	2–4 mm
Sabatieria	1B	2	Sp	cla	1–2 mm
Ascolaimus	1B	2	Cr	co	2–4 mm
Theristus	1B	2	Cr	co	>1 mm
Paramonohystera	1B	2	Cr	cla	1–2 mm
Promonhystera	1B	2	Cr	e/f	1–2 mm
Desmolaimus	1B	2	Cr	cla	1–2 mm
Odontophora	1B	2	Cr	co	2–4 mm
Steineria	1B	2	Cr	cla	1–2 mm
Terschellingia	1A	3	Cr	e/f	2–4 mm
Anticoma	1A	2	Pk	e/f	1–2 mm
Synonchiella	2B	4	Sp	e/f	2–4 mm
Viscosia	2B	3	Pk	cla	1–2 mm
Metoncholaimus	2B	3	Pk	cla	2–4 mm
Oncholaimus	2B	4	Pk	cla	2–4 mm
Oncholaimellus	2B	4	Pk	cla	> 4 mm
Bathyeurystomina	2B	4	Pk	e/f	> 4 mm
Marylynnia	2A	3	Sp	e/f	1–2 mm
Comesoma	2A	3	Sp	co	>1 mm
Prochromadorella	2A	2	Id	co	1–2 mm
Cyatholaimus	2A	3	Sp	co	2–4 mm
Paracomesoma	2A	2	Sp	cla	2–4 mm
Calomicrolaimus	2A	3	Sp	co	1–2 mm
Cobbia	2A	3	Cr	e/f	>1 mm
Desmodora	2A	2	REL	co	1–2 mm
Spirinia	2A	3	REL	co	2–4 mm

Table 2.
List and functional traits of nematodes genera identified in the Bizerte lagoon.

widely used ecological index is species richness, as it gives a clear and simple indication of the species number present in a sample. However, this measure strongly depends on the sample size or the environment studied [38]. Other diversity indices such as Shannon-Wiener and Pielou regularity as well as Index of Maturity [29] and Trophic Diversity [30], have been proposed and are regularly used to describe nematode assemblages in different environmental conditions [37]. The main advantage of diversity indices over the richness is that they give a better picture of the dominance of the species by considering the relative abundance of each taxon. Several studies investigated the impact of various environmental pollutants on benthic ecosystems using nematodes diversity indices to investigate their toxicity and harmful concentrations.

The results of nematodes exposure to metals during one month such as chromium have shown significant differences between univariate indices measures (most were decreased significantly). The responses of nematode species are characterized by the disappearance such *Leptonemella aphanothecae* which shows its high sensitivity to chromium, and an increasing of *Bathylaimus* species showing their high resistance [39]. In the same context, treatment of nematodes with nickel showed significant differences between nematode assemblages from control microcosms and those from treatments. Most univariates measures, including diversity and species richness, decreased significantly with the increasing of the metal concentrations. Results from multivariate analyses of the species abundance demonstrated that nematodes responses were highly variable: *Leptonemella aphanothecae* was considered sensitive because disappeared at all nickel concentrations tested; *Daptonema normandicum*, *Neochromadora trichophora*, and *Odontophora armata* which significantly increased at the high nickel concentration appeared to be "opportunistic" species at this concentration whereas *Oncholaimus campylocercoides* and *Bathylaimus capacosus* which increased at all nickel concentration seemed to be "nickel-resistant" [40]. Experiments of pesticide exposure such as permethrin have demonstrated that univariate indices were significantly different. The multivariate analyses revealed that nematode species *Pselionema* sp., *Prochromadorella neapolitana*, and *Spirinia gerlachi* were eliminated at the low dose and seemed to be intolerant to permethrin; *Trichotheristus mirabilis* and *Xyala striata*, which increased with increasing contamination levels, seemed to be 'opportunistic' and/or 'resistant' species [41]. Results of glyphosate treatment showed that Shannon-Wiener Diversity was reduced in all treatments. Species such as *Marylynnia stekhoveni* and *Microlaimus cyatholaimoides* decreased in all treatments and appeared to be "glyphosate sensitive" whereas *Paramicrolaimus spirulifer*, *Paracomesoma dubium*, *Metacomesoma punctatum*, *Terschellingia longicaudata*, and *Daptonema hirsutum* seemed to be "glyphosate resistant" species [42].

Exposed nematode communities to environmental levels of pharmaceutics compounds were also investigated. Thus, penicillin G exposure has demonstrated that diversity (H′), species richness (d), equitability (J), and the number of species (S) were decreased significantly and *Kraspedonema octogoniata* and *Paracomesoma dubium* were seemed to be sensitive species; *Oncholaimus campylocercoides* as "opportunistic", whereas, *Nannolaimoides decoratus* is "penicillin G resistant" species [43]. In other studies, Ciprofloxacin exposure caused a decrease in diversity index and nematode species were responded differently: *Odontophora villoti* was considered "sensitive," whereas *Metoncholaimus pristiurus* as "opportunistic" and *Paramonohystera pilosa*, appeared "tolerant" [23]. The exposure with collagen has induced a reduction in diversity indices. Nematodes species such as *Ptycholaimellus ponticus*, *Theristus modicus*, and *Kraspedonema reflectans*, were classified as "collagen-sensitive", *Sigmophoranema rufum*, *Lauratonema hospitum*, *Enoploides spiculohamatus*, and *Trichotheristus mirabilis*, were "collagen-tolerant" species [44]. The response of nematodes to polybrominated diphenyl ether (BDE-47) was also studied, and the results showed that all univariate indices (Species number (S); Shannon diversity index (H′); Margalef's species richness (d) were decreased and Pielou's evenness (J')) were significantly modified, and the species of *Terschellingia* were considered "BDE-47 sensitive," whereas *Metoncholaimus pristiurus* and *Paracomesoma dubium*, were "BDE-47 tolerant" (**Table 3**) [19].

Currently, some studies have started using the new assessment approach (functional traits analyses) to address the effects of various contaminants on nematode species. Among these studies, many experiments were conducted to evaluate the impacts of ciprofloxacin on nematodes functional traits evolution. A change in the nematofauna structure was registered and characterized by low values of the

Chemicals compounds	Ecological indices responses	Functional traits responses	References
Chromium	H′ ↓, D ↓, J' (ns), S (species number) ↓		[39]
Nickel	H′ ↓, D ↓, J' (ns), S ↓		[40]
Permethrin	H′ ↓, D ↓, J' ↓, S ↓		[41]
Glyphosate	H′ ↓, D ↓, J' ↓, S ↓		[42]
Penicillin G	H′ ↓, D ↓, J' ↓, S ↓		[43]
Collagen	H′ ↓, D ↓, J' ↓		[44]
Ciprofloxacin	H′ ↓, D ↓, J' ↑, S ↓	Lower taxonomic diversity All Functional traits were modified especially the tail shape.	[23, 45]
BDE-47	H′ ↓, D ↓, J' ↑, S ↓	• Restructuration of nematodes biological traits • Amphid shape was the most changed	[19, 46]
cd, PVC, and Mixture	cd (H′ ↓), PVC (H′ ↓), Mixture (H′ (ns))	• Restructuration of nematodes biological traits • Feeding diet and amphid shape and were the most modified	[47]
Separate and mixed PAHs (anthracene, pyrene, and benzo[a]pyrene)	D ↓	Life history and feeding diet were the most modified	[48]

Table 3.
List of studies using ecological indices and functional traits approach of nematodes for assessment of ecosystem quality.

taxonomic diversity. The nMDS second-stage ordination plots for matrices including nematode genera and biological traits showed that all attributes were modified and the tail shape was the closest to the generic structure [45]. Another study that treated the toxicity and the interactions between metals and plastic (Polyvinyl chloride) demonstrated that the single treatments was toxic for marine nematodes. However, the mixture of these pollutants has a lesser lethal impact compared to their separate effects. The nMDS second-stage ordination of inter-matrix rank correlations for matrices already mentioned showed that the proximate functional trait to the taxonomic responses was the amphid shape [47]. The response of meiobenthic nematode communities to the effect of the polybrominated diphenyl ether, BDE-47 has shown a decrease in taxonomic diversity and a modification in all biological trait abundance. Only three functional traits (body length, feeding group, and amphid shape), presented a clear difference between the untreated and treated microcosms. The nMDS second-stage ordination of inter-matrix rank correlations indicated that the amphideal shape was the most modified functional trait [46]. Finally, a study examining the single and binary PAHs (anthracene, pyrene, and benzo[a]pyrene) toxicity on marine nematodes have demonstrated that the single or mixtures treatments exhibited restructuring of trophic diversity with an increase of epigrowth-feeders abundance. The nMDS second stage ordination of inter-matrix rank correlations showed that the feeding diet and life history were the most modified functional traits after treatment with PAHs [48]. Although some studies have the approach proposed here to assess the environmental pollutants, there is still a lot of work to do to validate the use of the new approach to assess the quality of the environment influenced by different types of stress (**Table 3**).

4. Conclusions

The use of the approach based on the functional traits of marine free-living nematodes seems to be at present more relevant than the classical ecological index analysis methods used to detect changes quality of aquatic ecosystems. This new approach has made it possible to provide additional ecological information on the nematode responses and then on ecosystem functioning [32]. All functional traits included Adult length; Amphid shape; Tail shape; Life history; and Feeding type [9, 32–35] constituted a good approach to the determinate of the ecological status of the ecosystem.

Conflict of interest

The authors declare no conflict of interest.

Author details

Ahmed Nasri*, Patricia Aïssa, Hamouda Beyrem and Ezzeddine Mahmoudi
Laboratory of Environment Biomonitoring, Faculty of Sciences of Bizerta (FSB), University of Carthage, Bizerta, Tunisia

*Address all correspondence to: a7mednas@gmail.com

References

[1] Banks KE et al. Increased toxicity to *Ceriodaphnia dubia* in mixtures of atrazine and diazinon at environmentally realistic concentrations. Ecotoxicology and Environmental Safety. 2005; 60:28-36.

[2] Cheng JO, Liu KK, Ko FC. Environmental assessment of persistent organic pollutants in surface sediments of the Danshui River basin, Taipei, Taiwan. Environmental Science and Pollution Research. 2020; 27:44165-44176.

[3] Jeong H et al. Heavy metal pollution by road-deposited sediments and its contribution to total suspended solids in rainfall runoff from intensive industrial areas. Environmental Pollution. 2020; 265:115028.

[4] Peng Q et al. Pharmaceutically active compounds (PhACs) in surface sediments of the Jiaozhou Bay, north China. Environmental Pollution. 2020; 266:115245.

[5] Deng H et al. Microplastic pollution in water and sediment in a textile industrial area. Environmental Pollution. 2020; 258:113658.

[6] Bellasi A et al. Microplastic contamination in freshwater environments: A review, focusing on interactions with sediments and benthic organisms. Environments - MDPI 7:30. 2020.

[7] Ranke J. Persistence of antifouling agents in the marine biosphere. Environmental Science & Technology. 2002; 36:1539-1545.

[8] Fuentes I et al. Long-term trace element assessment after a mine spill: Pollution persistence and bioaccumulation in the trophic web. Environmental Pollution. 2020; 267:115406.

[9] Semprucci F et al. Do the morphological and functional traits of free-living marine nematodes mirror taxonomical diversity? Marine Environmental Research. 2018; 135:114-122.

[10] Boldina I, Beninger PG, Le Coz M. Effect of long-term mechanical perturbation on intertidal soft-bottom meiofaunal community spatial structure. Journal of Sea Research. 2014; 85:85-91.

[11] Sckratzberger M et al. Colonisation of various types of sediment by estuarine nematodes via lateral infaunal migration: A laboratory study. Marine Biology. 2004; 145:69-78.

[12] Nasri A et al. Trophic restructuring (Wieser 1953) of free-living nematode in marine sediment experimentally enriched to increasing doses of pharmaceutical penicillin G. Ecotoxicology. 2016; 25:1160-1169. https://doi.org/10.1007/s10646-016-1670-6

[13] Moens T, Bouillon S, Gallucci F. Dual stable isotope abundances unravel trophic position of estuarine nematodes. Journal of the Marine Biological Association of the United Kingdom. 2005; 85:1401-1407.

[14] Schratzberger M et al. The structure and taxonomic composition of sublittoral meiofauna assemblages as an indicator of the status of marine environments. Journal of the Marine Biological Association of the United Kingdom. 2000; 80:969-980.

[15] Nasri, A. Impact of penicillin G on the trophic diversity (Moens & Vincx, 1997) of marine nematode community: results from microcosm experiments. Cahiers de Biologie Marine. 2015; 56, 65-72.

[16] Coull BC, Chandler G. Pollution and meiofauna - field, laboratory, and mesocosm studies. Biology, Oceanography and Marine Biology. 1992, Corpus ID: 83596631.

[17] Austen MC, McEvoy AJ, Warwick RM. The specificity of meiobenthic community responses to different pollutants: Results from microcosm experiments. Marine Pollution Bulletin. 1994; 28:557-563.

[18] Schratzberger M et al. Effects of paint-derived tributyltin on structure of estuarine nematode assemblages in experimental microcosms. Journal of Experimental Marine Biology and Ecology. 2002; 272:217-235.

[19] Nasri A et al. Ecotoxicity of polybrominated diphenyl ether (BDE-47) on a meiobenthic community with special emphasis on nematodes: Taxonomic and trophic diversity assessment. Environmental Pollution. 2021; 277:116727.

[20] Schratzberger M, Warwick RM. Effects of the intensity and frequency of organic enrichment on two estuarine nematode communities. Marine Ecology Progress Series. 1998; 164:83-94.

[21] Nasri A et al. Using meiobenthic taxa, nematofauna biological traits, and bacterial abundance to assess the effects of the polybrominated diphenyl ethers compound: Case study of tetrabromo diphenyl ether BDE-47. Science of the Total Environment. 2021; 770:145251.

[22] Wang X, Liu X, Xu J. Distribution Patterns of Meiofauna Assemblages and Their Relationship with Environmental Factors of Deep Sea Adjacent to the Yap Trench, Western Pacific Ocean. Frontiers in Marine Science. 2019; 6:735.

[23] Nasri A et al. Chronic ecotoxicity of ciprofloxacin exposure on taxonomic diversity of a meiobenthic nematode community in microcosm experiments. Journal of King Saud University-Science - Elsevier. 2020; 32:1470-1475.

[24] Platt HM, Warwick RM, Furstenberg JP. Free-living Marine Nematodes. Part 1 British Enoplids. South African Journal of Zoology. 1985; 20:177-177.

[25] Bezerra TN, Decraemer W, Eisendle-Flockner U, Hodda M, Holovachov O, Leduc D, Miljutin D, Mokievsky V, Pena Santiago R, Sharma J, Smol N, Tchesunov A, Venekey V, Zhao Z, Vanreusel A, 2020. Nemys: world database of nematodes [WWW Document]. Accessed through world Regist. Mar. Species.

[26] Shannon CE, Weaver W. The Mathematical Theory of Communication. University of Illinois, Urbana, Illinois; 1949.

[27] Pielou EC. The measurement of diversity in different types of biological collections. Journal of Theoretical Biology. 1966; 13:131-144.

[28] Margalef, R. Information theory in ecology. General Systems. 1958; vol. 3, 36-71.

[29] Bongers T. The maturity index: an ecological measure of environmental disturbance based on nematode species composition. Oecologia. 1990; 83:14-19.

[30] Heip C, Herman R, Vincx M. Variability and productivity of méiobenthos in the Southern Bight of the North Sea. Rapport et procès-verbaux des réunions. Conseil international pour l'Exploration de la Mer. 1984; vol. 183, 507-521.

[31] Chalcraft DR, Resetarits WJ. Mapping Functional Similarity of Predators on the Basis of Trait Similarities. The American Naturalist. 2003; 162:390-402.

[32] Schratzberger M, Warr K, Rogers SI. Functional diversity of nematode

communities in the southwestern North Sea. Marine Environmental Research. 2007; 63:368-389.

[33] Thistle D, Lambshead PJD, Sherman KM. Nematode Tail Shape Groups Respond to Environmental Differences in the Deep Sea; 1995.

[34] Bongers T, Alkemade R, Yeates GW. Interpretation of disturbance-induced maturity decrease in marine nematode assemblages by means of the Maturity Index. Marine Ecology Progress Series. 1991; 76:135-142.

[35] Wieser W. Die Beziehung zwischen Mundhöhlengestalt, Ernäh rungsweiseund Vorkommen bei freilebenden marinen Nematoden. Arkiv. För. Zoology. 1953; 2:439-484.

[36] Clarke KR. Non-parametric multivariate analyses of changes in community structure. Australian Journal of Ecology. 1993; 18:117-143.

[37] Clarke K, Gorley R, Somerfield P, Warwick R. Change in marine communities: an approach to statistical analysis and interpretation. Primer-E Ltd, Plymouth, UK, 2014.

[38] Sanders HL. Marine Benthic Diversity: A Comparative Study. Am Nat. 1968; 102:243-282.

[39] Boufahja F et al. An assessment of the impact of chromium-amended sediment on a marine nematode assemblage using microcosm bioassays. Biological Trace Element Research. 2011; 142:242-255.

[40] Hedfi A et al. Effects of increasing levels of nickel contamination on structure of offshore nematode communities in experimental microcosms. Bulletin of Environmental Contamination and Toxicology. 2007; 79:345-349.

[41] Boufahja F et al. A microcosm experiment on the effects of permethrin on a free-living nematode assemblage. Nematology. 2011; 13:901-909.

[42] Salem FB et al. Impacts of Glyphosate on a Free-Living Marine Nematode Community: Results from Microcosm Experiments. Journal of Coastal Zone Management. 2016; 19, 434.

[43] Nasri A et al. Effects of increasing levels of pharmaceutical penicillin G contamination on structure of free living nematode communities in experimental microcosms. Environmental Toxicology and Pharmacology. 2015; 40:215-219.

[44] Allouche M et al. Laboratory bioassay exploring the effects of anti-aging skincare products on free-living marine nematodes: a case study of collagen. Environmental Science and Pollution Research. 2020; 27:11403-11412.

[45] Nasri A et al. Restructuring of a meiobenthic assemblage after sediment contamination with an antibacterial compound: Case study of ciprofloxacin. Ecotoxicology and Environmental Safety. 2020; 205:111084.

[46] Nasri A et al. Using meiobenthic taxa, nematofauna biological traits, and bacterial abundance to assess the effects of the polybrominated diphenyl ethers compound: Case study of tetrabromo diphenyl ether BDE-47. Science of the Total Environment. 2021; 770, 145251.

[47] Wakkaf T et al. The individual and combined effects of cadmium, polyvinyl chloride (PVC) microplastics and their polyalkylamines modified forms on meiobenthic features in a microcosm. Environmental Pollution. 2020 266:115263.

[48] Hedfi A et al. Nematode traits after separate and simultaneous exposure to Polycyclic Aromatic Hydrocarbons (anthracene, pyrene and benzo[a]pyrene) in closed and open microcosms. Environmental Pollution. 2021; 276:116759.